HIGHLIGHTS OF ASTRONOMY

Max. Alt. ☉ - 73.5°
(Summer)
(Winter) → 26.5°

HIGHLIGHTS OF ASTRONOMY

BY

WALTER BARTKY

PHOTOGRAPHS BY YERKES OBSERVATORY

 PHOENIX BOOKS
THE UNIVERSITY OF CHICAGO PRESS
CHICAGO AND LONDON

PHOENIX SCIENCE SERIES

TO F. R. M. AND W. D. M.
IN RECOGNITION OF THE GREAT DEBT I OWE
THEM, BOTH AS A STUDENT AND AS A FACULTY
MEMBER OF THE UNIVERSITY OF CHICAGO,
THIS BOOK IS DEDICATED

International Standard Book Number: 0-226-03840-8

THE UNIVERSITY OF CHICAGO PRESS, CHICAGO 60637
The University of Chicago Press, Ltd., London

Copyright 1935 by Walter Bartky. All rights reserved. Published November 1935. First Phoenix Edition 1961. Fourth Impression 1970. Printed in the United States of America.

AUTHOR'S PREFACE

WHEN it was suggested that a text of this type ought to be written for the astronomical portion of the Introductory General Course in the Physical Sciences, one of the four introductory courses under the new undergraduate curriculum at the University of Chicago, the "cons" were obvious. There was an abundance of literature on the subject: authoritative textbooks by noted astronomers and also fascinating popular treatises written by those same savants.

The "pros" were equally obvious. As is well known, the New Plan is designed to give the student a general educational background before he ventures to specialize in any particular field. Existing books on astronomy were found to be not altogether satisfactory for our purpose; either they were too lengthy or certain phases, which we regarded as essential, were omitted.

THE particular difficulty always encountered was the student's early discouragement when he could not locate for himself the stars and planets in the sky. To eliminate this obstacle without the introduction of mathematics, certain simple charts have been constructed. It is hoped that the ease with which many interesting problems can be solved by means of these charts will encourage the student and prolong his interest beyond the brief duration of the lectures. In the selection of topics I have been partly influenced by the many questions asked after public lectures, as well as by the queries of students, and I have endeavored to answer those demands by incorporating representative problems. At the same time I have attempted to present a logical development of the subject, starting with the earth and terminating with the sidereal universe.

MY OBLIGATIONS are many: To Mrs. Chichi Lasley for the attractive pen-and-ink sketches; to Yerkes Observatory for the astronomical photographs; to Professor Harvey B. Lemon for his pioneer work in *From Galileo to Cosmic Rays;* to Mrs. George S. Monk and Dr. Reginald G. Stephenson for numerous suggestions; to Director Philip Fox of the Adler Planetarium and Astronomical Museum for the welcome he has always extended to University of Chicago students and for the many opportunities he has afforded the author of demonstrating the Planetarium and thus obtaining closer contact with the lay student; to Mr. Donald P. Bean, manager of the University Press, for his unfailing patience; and to the staff of the University Press in general for encouragement, generous cooperation, and advice.

WALTER BARTKY

ECKHART HALL
UNIVERSITY OF CHICAGO
September 21, 1935

CONTENTS

FOREWORD	viii
1. THE EARTH	1
2. TIME	26
3. THE SKY	57
4. THE MOON	113
5. CELESTIAL MECHANICS	142
6. THE SOLAR SYSTEM	162
7. THE SIDEREAL UNIVERSE	225
CONCLUSION	258
BIBLIOGRAPHY	260
GLOSSARY OF DEFINITIONS	262
INDEX	269

FOREWORD

And if it please you, so; if not, why, so.
—SHAKESPEARE, *The Two Gentlemen of Verona*

WHILE driving along the seacoast at midnight, we pause and look out over the inky waters to behold in the distance a red and a green point of light and hundreds of tiny white lights. Some might say, "What a pretty sight" and pass on, but we gaze a little longer at the twinkling lights and they tell us that a passenger liner is off to sea. . . . . We imagine the activity on its decks as the stewards dash here and there on errands for the passengers; inside the salon there will be gaiety, and we try to picture the dancers. We wonder what foreign port this vessel will seek and whether there will be storms or smooth seas all the way. . . . . But we also cannot tarry. We take one last look at these dimming lights, bid a mental *bon voyage* and drive on.

Few there are who have not watched with awe the setting sun, or admired the rising moon, or gazed spellbound at the stars in a cloudless sky. We shall not be so foolish as to attempt to describe the beauty of these celestial objects. Our task is merely a very brief retelling of the story man has invented about these lights. With huge telescopes we concentrate their light and accurately measure

FOREWORD

their position. Aided by the results of multitudinous laboratory experiments and powerful mathematical analyses, we have built and continue to build up the universe around us.

THE earth we live upon is a huge sphere eight thousand miles in diameter, with a mass measured in billions of trillions of tons. Revolving monthly about the earth is its satellite the moon, a mere quarter of a million miles away from us. The earth and moon travel approximately in a circle ninety-three million miles in radius about the sun. This sun has a diameter of almost a million miles and a mass more than three hundred thousand times that of the earth. It is our chief source of light and energy—in fact, all other sources are negligible in comparison.

But our earth is only *one* planet in a small group of planets that travel approximately in circles about the sun and owe their light to it. The largest of these planets so far discovered is more than eleven times the earth in diameter; the smallest, less than half. Thousands of smaller objects, called "planetoids," ranging in size from a fraction of a mile to a few hundred miles, likewise revolve about our sun, along with myriads of tinier particles. If we add to the objects so far mentioned those bodies which terrified the ancients because of their huge volume, namely, the comets, we have what is called "the solar system." Thus our earth is one member of the solar system.

ARE there other solar systems, that is, are there other suns with planets similar to the earth revolving about them? We leave this for the individual to answer for himself. But astronomers are agreed that those points of light that we call stars *are* other suns, some very similar to our own sun, others millions of times greater in volume. The nearest of these suns is hundreds of thousands of times as far distant from us as our sun. That is why, even in our powerful telescopes, they appear as minute points of light. If any of these stars have planets, we can hardly hope to distinguish the tiny light they would reflect from the enormous radiation of the star. Current research indicates that our sun is only one sun in a group of billions of other suns called a "galaxy." But this is not all; present-day observations point out that there are millions of such galaxies within the reach of our telescopes!

No matter how skeptical one may be of the details of this picture man has made of these points of light, nevertheless it seems fitting that the immensity of the heavens should be comparable with their majesty.

CHAPTER 1

THE EARTH

> And we wonder—how we wonder!—
> What on earth the *earth* can be!
> —*The Mikado* (with apologies)

O**N CONSULTING** *Funk and Wagnalls Practical Standard Dictionary*, we read, "EARTH, n. 1. The globe on which we dwell, considered as a whole: distinguished (1) from other heavenly bodies and (2) from the abode of departed spirits"—and then follows about a dozen more interpretations of the topic of this chapter. Now, as one of our standard references declares, in its very first definition, that the earth is a heavenly body peculiar in the fact that we live upon it, and as heavenly, that is to say "astronomical," bodies are the subject of astronomy, it behooves us to consider our earth in any study of the science, for it is the heavenly body about which we know most.

It is a widely accepted fact that, as stated in the foregoing definition, the earth is a globe or sphere; and we generally take this statement at its face value. As we look around, however, our eyes give the lie to our intelligence, for the earth does seem to be flat. Vague memories loom out of the fog of early school days—some one or other maintained his belief in the earth's sphericity and this belief resulted in the discovery of America; someone else was credited with circumnavigating the globe. When we objected that we could travel in a circle on a flat earth and arrive back at our starting-point, another so-called proof of the earth's sphericity was offered, namely, that the hull of a ship disappeared while the mast was yet visible. This, however, remained unconvincing, for many of us had never seen a sailing-ship except in pictures; but in order to avoid tiresome argument, we agreed that this was so and that the earth was indeed round. In our modern age, however, many time-hallowed fallacies have been exposed; so let us adopt a more scientific attitude, charter an aeroplane, and leave the earth's surface "for to admire and for to see, for to be'old this world so wide," and, principally, to determine its shape. We select the Great Lakes region for our flight, the large number of cities concentrated in this district being easily located landmarks.

A THOUSAND feet above the crowded streets of Chicago, we trace highways and railroad tracks until they disappear in Michigan City, Elgin, Waukegan, or other cities forty miles away; ascending three thousand feet more we see, on our northern horizon, Lake Shore Drive entering Milwaukee. When we reach seven miles above Chicago, Detroit looms up on our eastern horizon, and, as the earth's shadow finally falls on the motor city, we know that for our present position the sun will remain above the horizon for more than half an hour.

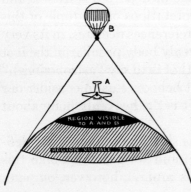

Cruising around during the long hours of the night, we wonder why it is that, the greater our height, the greater the portion of the earth that is visible. Does our vision improve with altitude? We check our results

THE EARTH

with powerful binoculars and telescopes and reject this theory. We draw diagrams of flat earths and curved earths and finally come to the conclusion that the earth *is* curved, for it is *curvature* that diminishes the region above our horizon at lower altitudes.

At considerable heights the horizon appears as a circle with a center that is directly below us; and, according to a geometrician in our party, if this is the case all over this "globe" of ours, then the "globe" must be a sphere. Perhaps here we have unwittingly hit upon the reason for the many altitude, endurance, trans-continental, and trans-oceanic flights. Following the example of Columbus, they too are proving in their own fashion what had been expounded by Pythagoras (sixth century B.C.)—the earth is spherical!

Our student of geometry informs us that, given our height and the maximum distance visible on the surface of earth, the diameter of the earth can be derived from the relation:

(Height) × (diameter of earth) = (square of distance visible).

Let us ignore this formula, as the distance visible could not be accurately measured. Instead, from our altitude of seven miles, we will watch the sun rise. At the very instant that the sun is com-

pletely above our horizon we "bail out"; and, as we fall, the sun appears to *set in the east*. When our hurtling descent is sufficiently diminished by the drag of the parachute, we have the unusual experience of witnessing the second dawn of the same day, for the sun will once more rise in the east!

Back on earth again, we try to adjust our visualizations of everyday activities on a sphere. As we walk upright on this curved earth, we picture ourselves as continually changing our in-

clination. According to our view of things, people living directly "below" us must be standing on their heads; anyone standing a quarter the way around the globe would be at right angles to us; if two people are standing erect and make an angle of 45° with each other, then they are separated by a distance equal to one-eighth the circumference of the earth. In fact, the angle between two people gives the fraction of the earth's circumference that they are apart; and, if we know the distance between two individuals on the surface of the earth and the inclination of the one to the other, the circumference of the earth may be found.

Here we have a method of determining the earth's radius that sounds a trifle more practical than boring a tunnel down through its crust to the center and stretching a tapeline from center to surface. All that we require is two upright men stationed at different portions of the globe; but, to obtain any reliable measurements, they must be rather far apart—so far that the one cannot see the other. The problem of finding the distance between them is easily solved by driving an automobile from one to the other, taking the most direct route, of course, and reading the mileage as shown by the speedometer. There remains only the determination of the angle between the two men; and we ask them to look overhead, for now the stars must be brought into the problem.

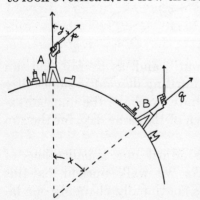
To illustrate the method, let us suppose that we have stationed one observer at Chicago and the other at New Orleans and arranged that they both observe the stars at exactly the same time of night, keeping in communication with each other by long-distance telephone. Observer A at Chicago sees one group of stars directly overhead and observer B at New Orleans another group, for, because

$$\frac{AB \text{mi}}{\gamma° (\text{or } x)} = ? \text{mi}/{}_{1}° \times 360 = \text{Circum.} \oplus$$

THE EARTH

of the curvature of the earth, what is directly overhead for the one is not directly overhead for the other. A line drawn from A to the star that is directly overhead for B would be inclined as much to A as is B, i.e., angle x is equal to angle y, as shown in the accompanying figure. (You will note that the lines p and q are parallel, the reason being that any star observed is so many billions of times greater than terrestrial distance that the departure from parallelism defies detection.) In this way, without seeing B, A can find B's inclination to himself by measuring the angle y between the star directly overhead at Chicago and the star reported to him as being directly overhead at New Orleans. He will find this angle is approximately 12°. Along the most direct route, the distance between Chicago and New Orleans is about 830 mi. Summarizing the results—if we travel 830 mi. on the globe, our inclination changes 12° relative to our original position, or roughly 69 mi. for each degree. For 180° change in inclination, we must therefore travel about 12,500 mi. Such a change would place us on our heads relative to our original position, so that 12,500 mi. is halfway around the globe; "all the way around the earth," or its circumference, is therefore approximately 25,000 mi. To obtain the diameter of a circle, the circumference is divided by π (which is equal to 3.1416), so that, according to our observations, the earth's diameter is 7,950 mi., which makes the radius a little less than 4,000 mi.

THE previous example illustrates the fundamentals of the method actually employed in determining the size and shape of the earth. Eratosthenes of Cyrene (276–196 B.C.), following a similar procedure, is credited with measuring the earth's circumference, his error being less than 1 per cent. Though the modern method remains the same in principle, certain refinements have been added. First, surveyors measure distances with very high precision. Second, in order to locate the point directly overhead, a plumb line is employed, just as plumb lines are used in construction so that buildings will stand upright and will not incline toward their neighbors. If the change in direction of the plumb line were always the same for the same distance traveled wherever we were on the surface of the earth and whatever the direction traveled, then the earth would be a perfect sphere. But, with very careful measurements it is found that, as one travels north or south, the change in direction of the plumb line for a fixed distance is a trifle greater at the equator than at the polar regions, i.e., the curvature of the earth in a northerly or southerly direction is greatest at the equator and least at the poles; in other words, the earth is slightly flattened at the poles. From measurements of the type described, it is possible to find its exact size and shape; and, according to J. F. Hayford, the earth's dimensions are as given in the table herewith. From various checks of the measurements, it is estimated that the earth's radius is known with an error that does not exceed 50 meters, that is to say, the uncertainty is now only 1 part in 100,000.

	Kilometers	Miles
Equatorial radius...	6,378.388	3,963.34
Polar radius.......	6,356.909	3,949.99
Difference.....	21.479	13.35

In our attempt to emphasize the polar flattening, or OBLATENESS, of the earth, perhaps we might have exaggerated its relative magnitude. A cross-section through the poles would be essentially a circle, for, although the difference in the distances from center-of-earth-to-equator and center-of-earth-to-poles is 13 mi., we must compare this with the mean or average radius of the earth, which is 3,957 mi. The ratio of these two distances is only 1:297; and, if we constructed an exact model of the earth, letting 1 in. represent 2,000 mi., at first glance it would appear a sphere, for the polar diameter would be only .013 in. shorter than the equatorial; mere particles of dust would represent the highest mountains and fine scratches the deep canyons.

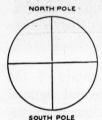

AT THIS point it might be asked, "Why is the earth oblate?" The answer is that the slight flattening at the poles and the bulging at the equator result from the centrifugal force produced by earth's rotation. This response is met with the query, "How do we know that the earth rotates?" and we now present our evidence for the rotation of the earth.

THE EARTH

We are all aware of the daily motions of the sun and moon in the sky, rising in the east and moving westward across the heavens to set below the western horizon. It was in the study of these motions of the sun and moon, along with those of the planets and stars, that not only astronomy but science itself began. The orderliness of the motions of the celestial objects above led man to seek order in the phenomena on the earth below. Looking to the north on a clear night in the middle northern latitudes, we see the stars describe circles about a point approximately midway between the point due

 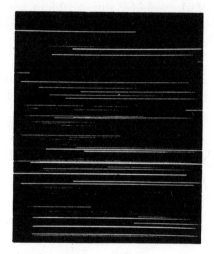

Star trails around the pole and at the equator

north on the horizon (technically known as the NORTH POINT) and the point overhead (or ZENITH). This point about which they seem to revolve is called the NORTH CELESTIAL POLE, and a photograph with a fixed camera shows the circular trails of the stars in this region.

Stars that rise due east travel in a semicircle to set due west 12 hr. later, while stars in the south seem to skim the horizon for only an hour or so. In fact, the stars as a whole seem to turn on a large sphere—the CELESTIAL SPHERE, as it is called—about an axis passing through the north celestial pole at one end and through another point below the horizon, the SOUTH CELESTIAL POLE. Naturally, man has asked if this be a real turning of the skies or merely an *apparent* motion due to an eastward rotation of the object upon which he lives. Our problem is to justify the latter theory.

OUR first thought is that an earth resting on nothing and spinning rapidly seems contrary to our everyday experiences. Nevertheless, we shall utilize the knowledge derived from those same experiences to prove the rotation of the earth. In proving the sphericity of the earth, we employed certain geometric properties of the sphere; our problem was geometric and we appealed to the laws of geometry. The point now in question involves motion, and we seek the laws of motion. Those of us who have studied physics, particularly the motion of terrestrial bodies, recall *Newton's Laws of Motion*. These laws were first definitely stated by Newton, although they were partly understood by Galileo and, before him, by the great artist-inventor Leonardo da Vinci.

I. Every body continues in a state of rest, or of uniform motion, in a straight line, unless it is compelled to change that state by a force.
II. The rate of change of velocity of a body is proportional to the force and inversely proportional to the mass, and the change takes place in the direction of the force.
III. To every action there is an equal and opposite reaction.

But explicit statements of laws, whether geometric, mechanical, or juridical, are tedious and difficult for most readers. We shall therefore try to reason entirely on the basis of our own terrestrial experiences.

IF THE daily motion of the stars is due wholly to the rotation of the earth, then any point on the earth's equator must be carried around its axis a distance of 25,000 mi. every 24 hr., or with a speed greater than 1,000 mi. per hr. This rotational velocity decreases with increase in latitude until it vanishes at the poles; at latitude 40° it is nearly 800 mi. per hr. How convenient it would be if we could deprive ourselves at will of this velocity; we could travel in about an hour from New York to Chicago. But when we glance at a formidable building just west of us, these advantages become somewhat less pleasing and rather more questionable—suppose we were to make an irretrievable and tragic mistake in will-power!

In wondering why our senses fail to reveal this motion, we recall vain attempts to sleep in an upper berth. As we rode along, we could not determine the speed of the train or whether it was going backward or forward. A stationary train, an earthquake to make it rattle, and a few cinders would do very well to bolster the illusion of the at times doubtful pleasures of railroad travel.

To while away the time in a Pullman, we construct a pendulum, simply suspending a heavy ball from the ceiling of the car by means of a baggage strap. As the train hastens along at 60 mi. per hr., we start our pendulum swinging at right angles to the car by pulling the ball to one side and releasing it. Back and forth the pendulum swings in a plane always at right angles to the car, provided our train continues at a uniform rate in a straight line. After long negotiations with the company and a lot of persuasion, we receive permission to repeat this same experiment with our car resting on a turntable, the pendulum swinging initially at right angles to the car. But suppose our car is being turned, will the pendulum move with the car? If the ball of the pendulum is heavy and the strap long, the effects of air currents and twisting of the strap will be negligible; there will be no appreciable force tending to change the plane in which the pendulum is swinging. Turning the car does not turn the pendulum; and, when the car has been rotated through 90°, the pendulum is seen to swing lengthwise. As the turntable slowly revolves the car clockwise, that is, in the direction in which the hands of a clock move, the pendulum appears to move counter-clockwise for the observers in the car.

TWELVE years prior to the appearance of the first Pullman, a French physicist, J. B. L. Foucault (1819–68), demonstrated the rotation of the earth by means of a huge swinging pendulum in the Pantheon at Paris. A globe 55 lb. in weight was attached to the ceiling by a metal thread 70 yd. long; and below the globe was a disk calibrated or divided so that the direction of the pendulum's swing was clearly indicated. When the pendulum was started swinging, the globe, when viewed from the south, slowly veered westward, proving that the earth rotated eastward. If the Foucault pendulum were mounted at the north pole, its performance would correspond to our Pullman-pendulum with the car on a turntable, for, as the north polar region rotates counterclockwise, the pendulum would appear to turn clockwise at the rate of 15° per hr. At the equator, the Foucault pendulum would behave as if it were mounted on a train traveling always in the same direction, and no turning would be observed. As the rate of veering increases with our latitude, its maximum being at the poles, in Chicago (latitude 42°) the Foucault pendulum appears to turn clockwise through 10° each hour, precisely the mathematically derived angle through which it should move at this particular latitude if the earth, rotating eastward, makes one rotation per day.

An extremely interesting instrument, which utilizes the earth's rotation, is the gyrocompass. Briefly, it is a heavy wheel spinning very rapidly about an axis which can move only in a horizontal plane. It can be shown by the laws of motion that, no matter where the axis of the wheel may point originally, because of the rotation of the earth it will eventually line itself so as to point to the earth's axis. In this manner the gyrocompass points to the *true north* in contrast to the magnetic compass, which depends upon the direction of the earth's magnetic field and which may therefore deviate considerably from the true north.

A CENTURY ago, we might have been overawed by the speed with which we are carried around the earth's axis; but, as we follow the rapid development of air transportation, we confidently look forward to the day when we shall truthfully say, "We left New York, the earth rolled by, and we arrived an hour later in Chicago." A recent speed record of over 400 mi. per hr. is actually greater than the rotational speed of a point on the arctic circle. How wonderful it will be to travel westward faster than the earth rotates eastward—the sun and stars rising in the west and setting in the east—dinner

THE EARTH

in London followed by luncheon in San Francisco 2 hr. later, at noon of the same day!

For the present, however, let us speculate on what would happen if we traveled westward at exactly the same speed as the earth rotates eastward. Devoid of the earth's rotation, the stars for us would neither rise nor set but would remain fixed in the sky. But what of the sun? At first sight, it, too, would appear to be fixed; but, as the days went by and we continued at our terrific pace, the sun

would seem to creep very slowly eastward in the midst of the stars. So slow would be this apparent motion that it would take the sun 1 yr. to return again to its original position among the stars.

UNFORTUNATELY, we must postpone to some future day our non-rotational flight. We come down to earth and from there try to establish this apparent eastward motion of the sun. Of course, during the daytime, the stars are not visible, for our atmosphere diffuses the light of the sun and thus prevents our seeing the stars during the hours the sun is above our horizon. The blue skies so beloved by the song-writers are due to the fact that sunlight is a blend of many different colors, and the particles of air, dust, and water vapor in the atmosphere scatter more blue light than any other color. When poets ascend to the greater altitudes of the stratosphere, they will probably rhapsodize about black skies studded with stars night and day. But we on earth must, for the present, locate the sun's position among the stars by noting just what group of stars is west of the sun before sunrise and what group is east of the sun at sunset.

Day by day, as our forefathers did thousands of years before us, we watch the slow eastward march of the sun among the stars. Every 24 hr. the sun moves over about twice its own diameter (1°

each day, or, more precisely, 360° in 365.256 days). A year passes; the sun has completed its circuit among the stars and we face the problem of accounting for this yearly motion. The hypothesis that man found most pleasing to his innate conceit was that the earth's center was stationary and that the sun actually traveled about the earth. Quite a natural hypothesis but wrong, alas! A few impious individuals, such as Aristarchus of Samos, who was born more than three hundred years before Christ, dared to arouse the righteous

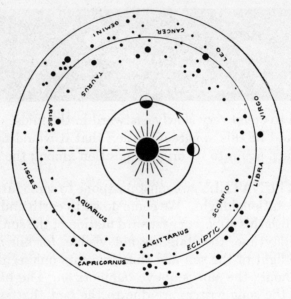

wrath of mankind by suggesting that the center of the abode of man was actually in motion and that this eastward journey of the sun was only an *apparent* motion! They maintained that the *earth* revolved yearly about the sun, which gave earthly observers the *illusion* of an eastward drift of the sun in the heavens.

UP TO a century prior to our Declaration of Independence, there were various ways of quieting the voices of those who persisted in their failure to appreciate the supremacy of this earth and the inhabitants thereof. If scientific or pseudo-scientific methods of persuasion failed to silence the scoffers, there was always

THE EARTH

the stake. Let us, however, confine our attention to the observational evidence which bolstered their belief in a stationary earth. Even the ancient Greeks realized that, if the earth were in motion about the sun, the nearby stars ought to appear to move with respect to the more remote stars. Every theatergoer has been forcibly made cognizant of this principle. Very often our vision of the moving-picture screen is obstructed by a massive head directly in front of us. Much as we might like to remove the obstruction entirely, we move our head to one side, and the nearby offending pate *appears* to move with respect to the far-distant screen. As we move our head from left to right, the heads of the patrons nearest us *appear* to move from right to left, relative to the more distant heads. But this phenomenon is not restricted to heads alone. If, as Aristarchus suggested, the earth circles the sun, moving from one side of the sun to the other, an analogous back-and-forth apparent motion should be produced in the closer stars relative to the more distant ones. His colleagues searched for this apparent yearly shift in the position of the nearby stars, or ANNUAL PARALLAX, as it is called; but they saw none. They concluded that the earth's center was stationary and that Aristarchus' theories should be abandoned.

Almost two thousand years pass and still the earth remains stationary—in scientific thought. Again dissenters appear. A Polish monk—Nicolaus Koppernik (1473–1543), or Copernicus, as he is better known—lived just long enough to see his infamous ideas of a revolving earth in print. Seventy years later his work was suppressed; but, during this interval, his ideas had taken firm root in the scientific world. During the troubled years of the Renaissance, many philosophers were persecuted for holding views which we now accept as commonplace. For instance, the Italian Giordano Bruno (1549–1600) went to the stake for upholding Copernicus' beliefs; Galileo (1564–1642) was also a suspect and only escaped a like fate by publicly retracting his arguments for a revolving earth.

However, it seems that about this time the natural inborn obstinacy of the human animal became more assertive; persecution actually appears to have fostered the idea of a moving earth. While our forefathers were busy colonizing America, European scientific

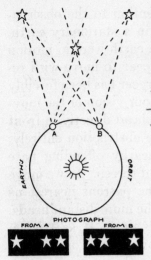

circles were not only accepting this theory but even had the temerity to offer evidence in favor of it. So convincing were their arguments that when parallax, this apparent motion of nearby stars with respect to the remote ones that *must* exist if the earth revolves, was finally discovered and proved in 1840, it came as no great surprise, and the news was received rather calmly.

To comprehend the reasons for the failure of man to observe the phenomena of parallax, you might try this to relieve the ennui of a tedious theatrical performance: Sit erect, close one eye, and move the other about one-tenth of an inch to the right or left. The apparent shift in the position of the individual in the seat directly in front of you will be somewhere in the region of a thousand times the annual parallactic motion of the nearest star in the heavens! And furthermore, this motion on the part of the star takes place over a period of several months! Notwithstanding the minuteness of the apparent shift, the change of the pattern formed by the stellar images can be accurately measured by photographing the sky through powerful telescopes at different times of the year and using microscopes in the examination of the resulting plates.

Another complexity entering into the problem is this—the stars, as well as our sun, are not stationary but are in motion. Fortunately, however, they move essentially in straight lines, indicating, as those who appreciate the laws of motion will readily understand, that the forces between stars are relatively small. Annual parallax imposes on this straight-line motion a back-and-forth motion with a period of 1 yr., so that the nearby stars appear to pursue a wavy path. To return to your theater of parallactic experiments—when you depart and the remaining patrons are moving in a dignified manner in straight lines down aisles and rows, you might likewise pursue with your feet a straight course but move your head periodically from one side to the other, observing this superimposition of motions.

THE EARTH

WE NEXT consider the shape of the path or ORBIT described by the earth in its motion about the sun. It is quite a simple matter to determine whether a train is approaching us or receding from us, by observing the change in its angular size. When approaching us, the angle made with the eye increases; receding, this angle decreases. We apply the same method to determine the variation, if any, in the distance between earth and sun. By actual measurement, we find that the sun's angular size varies very little during the year and that hence our "engine" remains more or less at the same distance from us. More accurate observations, however, indicate that the sun's angular diameter is 3 per cent greater in winter than in summer. Thus the earth is found to be at its minimum distance from the sun, or at PERIHELION, January 3 or 4 and at maximum distance, or APHELION, July 3 or 4, the distance at aphelion being 3 per cent greater than at perihelion.

The shape of the earth's orbit can be determined if we know the distance and direction of the earth for an observer on the sun for each day of the year. Fortunately, this information can be ascertained on earth, for the change in the angular size of the sun, as already pointed out, gives us the variation in distance. The position of the sun among the stars for earthlings is diametrically opposite the point that the earth would occupy for solarians, if there are any. The shape of the earth's orbit has been found to be an ellipse. Such a curve might easily be demonstrated by connecting two fixed pegs with a slack string. Pull on the string with a pencil point, moving this point about the pegs so as to keep the string always taut, and a 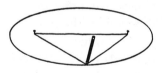 closed curve results—the ellipse. The position of the pegs is called the FOCI of the ellipse. If the distance between the foci is one-sixtieth the total length of the string joining them, then the resulting ellipse is of the same shape as the earth's orbit. (To complete the geometry of an ellipse—this ratio is called the ECCENTRICITY, while the length of the string is the MAJOR AXIS.) The sun is at one of the foci, which gives a rather asymmetrical, and perhaps unanticipated, arrange-

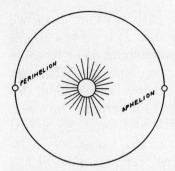

ment; there is no object at the other focus; nor, as we shall discover later, does this focus remain fixed.

We suggest that the reader draw with strings, pegs, and pencils an ellipse having the same shape as the earth's orbit. For his pains, he who complies will be rewarded with a curve that is essentially a circle; if he can detect any difference, then his construction must have been faulty, for this particular ellipse should depart from a circle by about one thirty-thousandth of its greatest diameter. To the eye, therefore, the earth's elliptical orbit *appears* a circle, with the sun off center.

We do not wish to imply by these geometrical constructions that there are any strings attached to the earth. It is gravity, that invisible force between masses, which causes us to drop and smash crockery and fall downstairs and which compels the earth to move as it does. If we could replace gravity by a steel cable with a diameter equal to that of the earth, it would snap like a thread under the stress of 20 tons per sq. in. necessary to hold the 6 billion trillion tons of matter in our earth in the course it pursues about the sun!

BUT WE are getting ahead of our story—we have not, as yet, revealed the actual size of the earth's orbit. This was intentional, for none of the observations so far mentioned enable us to determine the radius of the earth's orbit in terms of terrestrial units—miles or kilometers. The observed annual parallax of a star depends on the ratio of the distances of sun and star from earth, and the extremely small value obtained for this motion tells us that the nearest star is hundreds of thousands of times as far from us as is our sun. Later on we shall utilize measurements of parallax to determine the distances of the closer stars; but at this point we shall mention very briefly another proof of the revolution of the earth, a proof which also reveals the size of the earth's orbit.

We have all watched the rain strike the windows of a railroad car on a dull, wet day. When the train was stationary, the direction

THE EARTH

of the streaks left by the drops on the pane depended on the direction in which the rain was falling; but, to simplify the discussion, let us assume that the streaks were vertical when we stopped at a station. Then, when the train started and gathered speed, we found that the streaks became inclined to the vertical, the inclination increasing with the speed. It can be shown that, if the train traveled as fast as the raindrops were falling, the inclination of the streak would be 45°—halfway between the vertical and the horizontal; in fact, from the deviation of the streak from the vertical, we can determine the ratio of the velocity of the train to that of the rain. If the train backed up, the streak would be inclined the other way.

Light behaves as do our raindrops, the direction at which it appears to enter our telescope or our eye deviating from its true direction by an amount depending on the ratio of our velocity to the velocity of light. The velocity of light, however, is exceedingly high—186,284 mi. per sec., according to laboratory measurements of physicists*—and therefore in our daily activities we do not notice this phenomenon. But the velocity of the earth itself around the sun is so great that it produces a noticeable angular displacement in the position of all the stars, different from parallax in that the effect is independent of distance. This ABERRATION was noticed by Horrebow in 1659 and explained in 1727 by Bradley, the Astronomer Royal of England, in a manner analogous to our raindrops. He concluded that the earth traveled around the sun with a speed one ten-thousandth that of light—18.5 mi. per sec., according to modern measurements. Since it takes a year, or 31,500,000 sec., for the earth to complete its journey about the sun, the circumference of the earth's orbit must be 18.5×31,500,000 mi. If the earth's orbit were a circle, then its diameter would be this number divided by π (3.1416) or 186,000,000 mi.

THERE are several other very different methods of determining the size of the earth's orbit, and they all yield essentially the same result. The mean distance of the earth from the sun (i.e., one-

---

* Harvey Brace Lemon, *From Galileo to Cosmic Rays* (Chicago: University of Chicago Press, 1934), p. 401.

half of the sum of the maximum and minimum distances) is 92,900,000 mi. We call this distance the ASTRONOMICAL UNIT and find it a most convenient unit for expressing the smaller distances encountered in astronomy. For the benefit of those individuals who enjoy manipulating huge sums, we note that the Union Pacific streamline train, traveling 120 mi. per hr., day and night, would take 87 yr. to travel 1 astronomical unit; and the fare, at excursion rates of 2 cents per mile, would be roughly $1,860,000. Light travels this distance in 8 min., with no expenditure of energy, according to present-day concepts.

Those who are afflicted with periodic attacks of wanderlust might effect an inexpensive cure by reflecting on this celestial tour of the earth about the sun. We are somewhat boastful of man's accomplishments when we alight from a plane that has flown a mere thousand miles in 6 or 7 hr., and we may well feel proud of our feat when we compare our speed with that of a tortoise hastening along. We must remember, though, that the speed of the lowly tortoise bears the same ratio to our plane speed of a 150 mi. per hr. or better as does the plane to the orbital speed of the earth. On July 3 or 4 our orbital velocity is at its minimum value—65,500 mi. per hr. On passing aphelion, our speed increases as our distance from the sun diminishes, so that, when we reach perihelion half a year later, on January 3 or 4, we are 3,000,000 mi. closer to the sun and traveling 2,200 mi. per hr. faster—67,700 mi. per hr. Thereafter we slow down as our distance from the sun increases, until we again reach aphelion and a speed of 65,500 mi. per hr.

AS DEPENDABLE as day or night is Earth, our celestial train. In this motion, it obeys a definite law called the *Law of Areas:*

> The line joining the earth to the sun sweeps over equal areas in equal intervals of time.

If this statement means nothing to you, just imagine the earth as moving along the edge of an elliptical table-top. On the table, at one of the foci, is the sun. Let us suppose that the table is covered with a fine layer of dust. In order to remove the dust, we attach a long, thin brush a little more than 1 astronomical unit in length, so

THE EARTH

that one end pivots about the sun and the other end projects a little beyond the table and is dragged along by the earth. Then, according to the Law of Areas, the earth will sweep up the same area of dust each day, whatever the time of year. 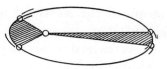 As the earth approaches perihelion, more of the brush extends beyond the table, less is used for sweeping, and the earth speeds up to maintain its quota of dust. Conversely, when approaching aphelion, less of the brush extends beyond the table, more is required for sweeping, and the earth need not hurry along so rapidly to maintain its quota.

WE HAVE all noticed how high the sun climbs in the sky in the early summer and how low it is, even at noon, when winter begins. In ages past, the sun was worshiped as the chief source of light and heat. Even now, all other supplies of energy pale into insignificance when compared to it. Let us, therefore, follow its daily path throughout the year. We shall record our observations in precisely the same primitive manner as did our forefathers, thousands of years ago: A straight stick is implanted upright on a level stretch of ground, and we watch its shadow closely. As the hours of morning pass, the shadow cast by the stick diminishes in length, until it is at its minimum at noon and the sun is at its highest in the sky. Each noon we note the length of the shadow. If we reside in the middle northern latitudes, we should find that, starting early in January, this noonday minimum length steadily decreases—very slowly at first, more rapidly in March and April, then slowing down again but still decreasing until the third week of June. As we approach June 21, the change in this minimum length becomes almost imperceptible, so that, when the sun is at its very highest point in the sky for the entire year, on June 21 or 22, we say it is the SUMMER SOLSTICE—from *sol*, "the sun," and *sto*, "stand." Thereafter, the length of the stick's shadow at noon increases, very slowly in July, more rapidly in September and October, and then slowing down as we approach the WINTER SOLSTICE, about December 22, where it stops increasing and starts to decrease. Thus the ancients determined the solstices; we seek their cause in the motion of the earth.

As we hasten on our long journey about the sun, the earth turns majestically on its own axis. The plane in which the earth's center moves is called the PLANE OF THE ECLIPTIC. The axis does not stand at right angles to the plane of the ecliptic but is inclined approximately 23°.5 away from the perpendicular; that is, the plane of the earth's equator makes an angle of about 23°.5 with the ecliptic. More precisely, the OBLIQUITY OF THE ECLIPTIC, as this angle is called, is 23°.45. In the earth's motion about the sun, we may regard its axis as pointing always in the same direction in space. Thus, during the

months preceding and following the summer solstice our north pole is tilted toward the sun; but a half-year later we are on the other side of the sun, and, since the north pole of the earth persists in pointing in its original direction in space, it is tilted away from the sun. The inclination being about 23°.5, it follows that at the time of the summer solstice the sun is seen directly overhead at Lat. 23°.5 N. (the Tropic of Cancer), while at the winter solstice the sun is overhead at 23°.5 S. (the Tropic of Capricorn).

To determine the maximum height of the sun at the solstices, we need but know our own latitude. Let us suppose that we are at 40° N. The sun is so very far away (almost 93,000,000 mi.) as compared to terrestrial distances that lines drawn from different points on the earth to the sun's center may conveniently be regarded as parallel. We have already pointed out that, as we travel along the

THE EARTH

curved surface of the earth, our individual inclination changes, so that a man standing upright at Lat. 40° N. makes an angle of 40° − 23°.5 = 16°.5 with a second man standing due south of the first at the Tropic of Cancer. If the sun is at the *zenith* for the second man, the zenith being the point in the sky directly overhead, it follows that it will be 16°.5 below zenith for the man at Lat. 40° N. By a similar argument, we find that at the winter solstice, which occurs December 21 or 22, the sun at its highest is 16°.5 + (2 × 23°.5) = 63°.5 away from the zenith of Lat. 40° N., the total variation in the angular height of the sun at noon throughout the year being 2 × 23°.5 or 47°. For those of you who have chanced to observe the variation in the length of the shadow you cast at different seasons on a level stretch of ground, we mention that for Lat. 40° N. the length of your shortest shadow on June 22 would be less than a third your height; on December 22 it would be more than twice.

WE NOW turn our attention to a more important shadow—that cast by the earth upon itself. We all know that essentially half of the earth is always in the sunlight and the other half in darkness. But in general on a given day the northern and southern hemispheres do not share equally in the sunlight striking the earth. When the north pole is directed toward the sun, the northern hemisphere receives the greater portion, which, at the time of the summer solstice, amounts to as much as 70 per cent of total energy intercepted. On this date, June 21 or 22, the sun's rays strike 23°.5 beyond the north pole of the earth, so that in Lat. 66°.5–90° N. the sun is visible throughout the entire day—the Land of the Midnight Sun. From Lat. 66°.5 N. (the arctic circle) south to the equator, the duration of sunlight decreases gradually from 24 hr. to 12 hr. South of the equator, night exceeds day; and the duration of daylight decreases from the equator to the antarctic circle, Lat. 66°.5 S. Since the south pole is tilted 23°.5 away from the sun, Lat. 66°.5–90° S. are in the earth's shadow the entire 24 hr. during the period in which the same latitudes north are enjoying continuous sunlight. Six months later conditions are reversed. On December 22, 70 per cent of the energy allotted to the earth falls south of the equator; south of the antarctic circle is then the Land of the Midnight Sun; night is longer than day in the northern hemisphere; the arctic circle is enduring its period of darkness.

We are all familiar with the consequences of this inequality in the daily amount of heat received by a given hemisphere. Those of us who suffer the tortures of the inferno in the middle northern latitudes during the days following the summer solstice know too well how the sun terminates our troubled slumbers by rising, far north of east, much too early in the morning. At latitude 40° N. it rises $7\frac{1}{2}$ hr. before noon, then climbs up higher and still higher in the sky; its rays beat down upon us, heating both us and the surrounding territory for 15 hr. In the short 9-hr. night, we cannot cool off, so that, when the sun commences its merciless repeat performance, we still have some heat left from the previous day. Day by day we accumulate more and more heat, and, on the whole, our temperature steadily rises. It is true, of course, that following the summer solstice the days grow shorter, the maximum height of the sun diminishes, and, consequently, the heat received from the sun decreases. This diminution of radiation received, however, is scarcely appreciable for two weeks or so following the summer solstice, and the accumulation continues through July and August, depending upon atmospheric conditions and the nature of the surrounding territory. That is why we say that summer only begins at the time of the summer solstice, the day we receive our maximum amount of heat. Finally, however, the earth's radiation of heat exceeds that received from the sun, and, on the whole, our average temperature steadily drops; and this cooling continues long past the winter solstice, the date at which we in the northern hemisphere receive our minimum amount of heat and the date regarded as the official beginning of winter.

There is no necessity for us to dwell upon the miseries we endure in winter; let it suffice that we have traced the fundamental cause of the seasons to the inclination of the earth's axis relative to the plane in which we move about the sun. The subject of weather, considered a safe topic by all guides to the art of polite conversation, falls outside the scope of astronomy. We shall not take into account

THE EARTH

the rôle of atmospheric currents and local topography in the determination of climate. Astronomical calculations merely tell us the quantity of heat supplied by our solar furnace. For example, we know that the north and south poles, on June 21 and December 21, respectively, receive more heat in 24 hr. than any other point on the globe. What erroneous conclusions we might draw as to their climate if we neglected to take into account the frigid effects of their enormous supply of ice!

Following December 21, the winter solstice for the northern hemisphere, the duration of daylight increases; but, as this date is the summer solstice of the southern hemisphere, the duration of daylight decreases in those regions south of the equator. On or about March 21 the sun is seen directly overhead at the earth's equator, and day and night are equal over the entire globe. We say, then, that the sun is at one of the EQUINOXES (*aequus*, "equal," and *nox*, "night")—in this case it is the VERNAL EQUINOX, for this event is taken as marking the beginning of spring. The vernal equinox plays a very important rôle in our schemes for locating celestial objects. It should be regarded as a definite point among the stars; and, since the sun appears to move annually among the stars, the vernal equinox is the particular position occupied on or about March 21 by the sun's center when it is directly over the earth's equator. Day and night are again equal when the sun is once more directly overhead at the equator, or at the AUTUMNAL EQUINOX, which occurs September 22 or 23, the official beginning of autumn.

IF ON a diagram of the earth's elliptical orbit we located the position of the earth at the two solstices and the two equinoxes and drew a line through the first two points and another line through the second pair, we would find that the lines intersected at the sun and were at right angles to each other; that is, in passing from the official beginning of one season to the next, the line joining earth to sun always turns through an angle of 90°. But we have seen that, according to the Law of Areas, this line must sweep over equal areas in equal intervals of time; and hence its angular rate of turning is not constant. On or about January 4 the earth is at perihelion (minimum distance from the sun), and this imaginary line is therefore turning at its fastest. Consequently, winter is the shortest season in the northern hemisphere. To emphasize the effect of the ellipticity of the earth's orbit, we suggest that the reader consult a calendar and practice a little subtraction. He will find that, for the northern hemisphere,

> Summer is 4.5 days longer than winter,
> Spring is 3.8 days longer than winter,
> Autumn is 0.9 days longer than winter.

Of course, in the southern hemisphere, the seasons are just the reverse, summer being the shortest and winter the longest. It is interesting to note that, since the earth is closest to the sun in January, the southern hemisphere receives almost 7 per cent more heat on December 21 than does the northern hemisphere on June 21. But summer, for latitudes south of the equator, is shorter than for those north; and it is found that the total amount of heat received in northern and southern hemispheres is equal for this season. The same result applies to all corresponding seasons in both hemispheres. Nevertheless, the yearly variation in the amount of heat received daily for southern latitudes is greater than for the corresponding northern latitudes. We might therefore expect greater extremes in temperature in the southern hemisphere if we forgot about the greater abundance of water in this half of the globe, which naturally tends to reduce the variation in temperature. Let us agree to postpone this question for about 12,000 yr., for then, as we shall find later, conditions, as far as solar heat is concerned, will be interchanged in the two hemispheres. At this time the inhabitants of the northern portion of the globe will be closest to the sun in summer.

WE PAUSE a moment, in thought but not in space, to reflect upon this huge whirling sphere which is the abode of man. We could add to our list other motions besides rotation and revolution, motions participated in by both earth and sun; but let us be content for the present by combining merely rotation and revolution. Placing ourselves in space about 1 astronomical unit from the

THE EARTH

sun, directly over the plane of the ecliptic and the northern hemisphere, we see a tiny point, the earth, circling the sun in the opposite direction to that traveled by the hands of a clock. We view the earth with a powerful telescope and note that it also turns counterclockwise on its axis. We focus our telescope and attention on some spot in the tropical or middle northern latitudes. The sun's rays divide the globe into two hemispheres—one in sunlight, one in darkness. Likewise the motion of the earth about the sun yields another division—the "forward" half and the "rear" half. Daily rotation carries our spot from sunlight to darkness and from the "forward" to the "rear" side of earth. Since the earth's center moves approximately at right angles to the line joining earth and sun, we find that, when the sunlight appears on our spot, it is on the "forward" half of the globe; at sunset it is on the "rear." Weary with our long journey through space, we return to earth and gaze longingly on our beds, trying to realize that, relative to the sun, they are descending at sunset about a 1,000 mi. per min., to rise again with this same terrific speed 12 hr. later.

CHAPTER 2

TIME

> I saw Eternity the other night . . . .
> . . . . Time in hours, days and years,
> Driven by the spheres, like a vast shadow moved;
> In which the World and all her train were hurled.
> —HENRY VAUGHAN (1621-95)

A VOICE is heard over the radio—"At the sound of the gong the time will be 8:30 P.M. Central Standard Time." We pause to adjust our watches and to reflect upon the properties of this important physical concept—time. So vital is it to man that we interrupt a program of dance music or a tooth-paste panegyric to broadcast its value. What *is* time? Perhaps we should answer that time is time, for the word itself conveys more to us than any other term we might use in attempting to define it. The same might be said of space. These two undefined concepts, space and time, play very similar rôles in our lives.

The earth upon which we live constitutes a material world; and in it we deal with events—the collision of two or more automobiles,

TIME

a gangster "liquidation," a prize fight, a "society" wedding, and so forth—see any newspaper for further examples. The *blasé* will pardon us for regarding all these occurrences as "events," for here we are utilizing the word in its broadest sense to include anything that happens. In reporting or cataloguing such incidents, we not only locate their position in space but also state the time of their occurrence. Thus, space and time are both means of ordering events.

ALTHOUGH we do not define space or time, we do describe their measurement. According to elementary notions of geometry, we can move line segments around in space and still preserve their length. We utilize this principle when we obtain the distance between two points by laying down a yardstick repeatedly. Thus our measurement of space is based on the laws of geometry. Let us next consider the measurement of time. Since time is a means of ordering events, we might attempt to gauge it by the number of experiences of equal intensity that occur to us in a given interval, and thus arrive at a sort of psychological time. As a concrete example, we might reckon time by the number of pages of a theme that we could write or the number of blocks that we walked. Basing our time on the amount of work accomplished might be satisfactory for the individual, but this would result in a different system for each and every mortal and would be most confusing, to say the least. We therefore seek an impersonal basis.

Just as the laws of geometry led to a practical basis for measuring space, so the laws of motion lead to a method of determining equal time intervals. According to the First Law of Motion, a body in motion and subjected to no external forces will pass over equal distances in equal intervals of time. Since equal distances are obtainable according to the laws of geometry, that is, practically by means of a yardstick, we may define two intervals of time as equal if a moving body subjected to no exterior forces passes over equal distances in them. To illustrate this principle, imagine that we have a long level table-top. On this table-top we slide a small block, and so smooth are they both that the table offers no resistance to the motion of the block. If we mark off equal distances along the table-top, then, as it slides along, the block would move from one marking to the next in equal intervals of time. But our table-top clock exists only in our imagination, for we cannot escape friction.

Nevertheless, all our present-day time-devices are reconcilable with the laws of motion. The movement of the hands of a pendulum clock or of a watch can be analyzed by the methods of Newtonian mechanics and shown to be approximately uniform, that is, the minute or the second hand sweeps over equal angles in equal intervals of time. However, the most accurate and practical timepiece so far discovered is the earth itself. According to the laws of motion, a rotating body, subject to no exterior forces and to no changes in the distribution of its mass, rotates at a uniform angular rate. The earth satisfies those conditions so closely that, to date, no man-made clock has detected any departure from a uniform angular rate of spin; hence we take the turning of the earth as our standard of time. Two intervals of time are equal if, during them, the earth rotates through equal angles.

TO MEASURE time we watch the stars, for it is the eastward turning of the earth that causes the stars to appear to travel westward. When we look up at starry skies, it seems as though we were at the center of a huge sphere, the *celestial sphere*, studded with points of light; as the hours pass, the sphere seems to turn slowly about us, carrying the stars along with it. The angular rate and the axis of rotation for this imaginary sphere correspond to those of the earth. If we extended the earth's axis indefinitely in both directions, it would pierce the celestial sphere at two points—the *north and south celestial poles*. At the north pole of the earth, the north celestial pole would be seen directly overhead, making an angle of 90° with our horizon; but, as we travel southward along the curved surface of the earth, the angular height of the north celestial pole decreases by an amount equal to our change in latitude. At the north pole, Lat. 90° N., the angular height of the north celestial pole above the horizon is 90°; at Lat. 80° N., 10° south of the north pole, it is 90°−10°, or 80°. In fact, we can write a very simple and important rule for determining the angular height, or, as it is technically designated, the ALTITUDE, for the north celestial pole, namely:

Altitude of north celestial pole is numerically equal to the latitude of observer, provided he is in the northern hemisphere.

After memorizing this rule, we suggest that the reader select a clear night, look due north at an angular height equal to his latitude, and locate the north celestial pole, checking his observation by noting that all the stars appear to circle this point. Let him now trace a semicircle in the sky, which starts at the north celestial pole, continues upward to pass through his own zenith (the point directly overhead), then downward to the *south point* (the point due south on his horizon), to terminate below his horizon at the south celestial pole, the latter being as far below his horizon as the north celestial pole is above. In this fashion the reader will find his MERIDIAN in the sky. Now the word "meridian" is also encountered in geography, and there it is used to denote semicircles on the earth terminating at the poles. The *celestial meridian* for each observer is directly over the terrestrial meridian upon which he is standing.

All through the night we see stars rising in the east, climbing upward in the sky to reach the meridian, then descending to set in the west. We select a particular star and watch it cross the meridian. The next night it repeats the performance, and the interval of time from one TRANSIT of the star across the meridian to the next transit equals the length of time it takes the earth to make one complete rotation. This interval is called the SIDEREAL DAY. It is divided into *24 sidereal hours* numbered from 0 to 23, each sidereal hour is divided into *60 sidereal minutes*, and each sidereal minute into *60 sidereal seconds*. The reason for the introduction and reiteration of the adjective "sidereal" (it means "pertaining to the stars") will be apparent later.

THE question now arises, When should we start counting sidereal time? We might select some exceptionally bright star and say that, when this particular star crosses our meridian, it is 0 hr. sidereal time. However, our *alpha* for sidereal time is the transit of the vernal equinox across the meridian, which, as we have seen, is a point among the stars, namely, that point occupied by the sun's center at the official beginning of spring, on or about March 21. It is a great pity that, on this date, we cannot climb up to the celestial sphere and attach to the stars right at the sun's position a neon sign bearing the words "Vernal Equinox." Then, as the days passed and the sun continued away from the vernal equinox on its apparent eastward journey among the stars, we could still quickly find the desired point. When the words "Vernal Equinox" flashed on our meridian, then we would know that it was 0 hr. sidereal or star time. Nevertheless, the determination of sidereal time by the stars is almost as simple. Each star crosses our meridian at a definite sidereal time, which varies by only a few seconds during the year, this variation being due to aberration, a slight motion of the equinoxes and of the star itself—factors that need not concern us unless the highest degree of accuracy is desired. The sidereal time at which a star is on the meridian is called its RIGHT ASCENSION. Thus the right ascension of Sirius, the brightest star in our sky, is 6 hr. 41 min.; that is, at 6 hr. 41 min. sidereal time, Sirius is due on our meridian and will be there at that time, despite rain, hail, or snow.

All we need to determine sidereal time is a time-table of the stars and a clear night. The former is more easily obtained than the latter, for the United States government publishes a book of extreme importance to astronomers as well as to navigators, known as *The American Ephemeris and Nautical Almanac*, which, among other vital statistics, gives the right ascension of certain stars to a hundredth of a second. These stars are very frequently referred to as "clock" stars; and, by watching the transit of such a star across the meridian through a suitable telescope (a telescope often used is one which is mounted so that it can be pointed only at the meridian, and is called a MERIDIAN CIRCLE), the astronomer records his sidereal time to a hundredth of a second.

Although all other systems of time are computed from observations of star time, nevertheless time, as reported by such services as Western Union, radio, telephone, etc., is not sidereal

TIME

time. Our normal daily activities conform more or less with the *sun's* position in the sky, but this dictator of our earthly pursuits does not maintain a fixed schedule according to the *stars*. It is our problem to find its sidereal time-table.

We note first that the vernal equinox is always "on time," sidereally speaking, crossing the meridian at 0 hr. according to our sidereal clock; if it does not, then we adjust the clock. On or about March 21 the sun is at the vernal equinox and hence crosses our meridian at 0 hr. sidereal time; that is, the right ascension of the sun is about 0 hr. on March 21. But the sun does not remain at the vernal equinox, for the yearly revolution of the earth causes the sun to appear to move eastward among the stars and depart from that point. In one year the sun completes its apparent circle among the stars, moving at an average rate of 30° per month. On April 21 we find the sun approximately 30° east of the equinox, and, since it takes 24 hr. for the skies to turn through 360° (15° per hr.), it crosses the meridian 2 hr. after the vernal equinox; in other words, on April 21 the right ascension of the sun is about 2 hr. On May 21 the sun is approximately 60° away from the vernal equinox, and hence its right ascension is about 4 hr. We may now write a simple rule for the sidereal time schedule of the sun:

Right ascension of sun in hours = 2 × number of months after March 21.

Month by month, the right ascension of the sun increases about 2 hr., until it is once more at the vernal equinox; each day it changes by about 4 min. For example, if on March 21 the sun's center crosses our meridian at 0 hr. sidereal time, on March 22 it will cross at 0 hr. 4 min., according to our sidereal clock. In other words, the sun runs slow according to the stars, the solar day or interval between successive transits of the sun across the meridian being about 4 min. longer than the sidereal day.

Let us consider what would happen were we to regulate our affairs by sidereal time. On March 21, at high noon according to the sun, sidereal time is about 0 hr. One month later, sidereal time at noon is 2 hr.; two months later it is 4 hr.; and so on. If we had luncheon always at 0 hr. sidereal time, our repast each month would occur 2 hr. earlier with respect to the position of the sun until, on September 22, we would have luncheon at midnight, dinner at sunrise, and breakfast at sunset. Furthermore, in the course of a year we would gain three extra meals, for the number of sidereal days in the year is one more than the number of ordinary or, rather, solar days.

BUT custom has decreed that we base our time on the sun, and therefore we seek a satisfactory division of the solar day. We might adopt the scheme of the ancient Egyptians and Greeks and divide the period of light into twelve equal hours and the period of darkness into another set of twelve hours. With a six-hour working day, this mode of division would be quite agreeable in the northern latitudes during December and January; but, with the approach of summer, labor troubles would commence. In the thirteenth century, an Arabian mathematician suggested that the interval from one transit of the sun across the meridian to the next be divided into twenty-four equal hours and, some two centuries later, his proposal was generally adopted. The instant the sun's center is on our meridian we call APPARENT NOON, and the number of hours and fractions thereof expressed in minutes and seconds which have elapsed since apparent noon is known as APPARENT TIME. Other terminologies are often employed for apparent time, namely, TRUE SOLAR TIME and THE HOUR ANGLE OF THE SUN.

Those relics of the past—the sand glasses; the clepsydrae, or water clocks; and the elaborate sundials with which man used to measure apparent time or the hour angle of the sun—have been relegated to the museums as curios. Sundials, of course, are still encountered even in modern gardens or built into the walls of houses; but we now look upon them as merely decorative, and we would as soon use them for timepieces as we would measure the lapse of the hours by repeating certain psalms, as did the monks of old. We live in an age in which the sun has been discarded as a timepiece, for the solar days vary in length. This is not the fault of the sun, for the blame, if blame there be, should be placed on the earth. If the earth moved uniformly about the sun in the plane of the equator, then the sun's apparent angular motion eastward among the stars would be constant; and the difference between the solar day and the sidereal day, due to the annual apparent motion of the sun, would be the same throughout the year. But the angular speed of the earth varies throughout the year, and this variation is reflected in the sun's apparent motion. Again, at the time of the equinoxes, the sun appears to cross our equator, and this northerly or southerly motion is made at the expense of the eastward motion. On combining these two factors, it is found that the difference—solar day minus sidereal day—varies from 4 min. 26 sec., the longest solar day (about December 22), to 3 min. 35 sec., the shortest solar day (September 17). In other words, our sun is always slow according to our assumed perfect sidereal clock; around Christmas, as we conclude in childhood, its speed is even less than it is during the end of our summer vacation.

SINCE the solar day is more desirable for everyday purposes than the sidereal, we proceed to repair our solar clock. First we add up all the values of the solar day throughout the year and strike an average to obtain what is called the MEAN SOLAR DAY. Then we divide the mean solar day into hours, minutes, and seconds of mean solar time. Because there is one more sidereal day in the year than there are mean solar days, the mean solar day, hour, and second are a trifle longer than the corresponding intervals of sidereal time. Thus,

$$\begin{aligned} 1 \text{ sidereal day} &= 24^h \text{ sidereal time} \\ &= 23^h 56^m 4^s.091 \text{ of mean solar time.} \\ 1 \text{ mean solar day} &= 24^h \text{ of mean solar time} \\ &= 24^h 3^m 56^s.55 \text{ of sidereal time.} \end{aligned}$$

Watches and clocks which are used for ordinary purposes should tick off mean solar seconds; but, if your watch runs about 4 min. fast each day, do not discard it immediately, for it might be indicating perfect sidereal time. The bright student will note that, if the direction of the earth's rotation were reversed and everything else remained unaltered, we would have sunset and then sunrise rather than sunrise and then sunset, and, in addition, 2 more days in the year to fit into our calendar; furthermore, our watches would run faster by 8 min. each day!

IT MIGHT seem that we have overemphasized the variation in the length of the solar days, for, at most, the solar day departs from the mean by half a minute. This means that, during the latter portion of December, the sun is "running slow," according to our mean solar clocks, by half a minute each day, and these half-minutes will accumulate. If your watch were to run slow half a minute each day, before very long you would be compelled to make a considerable adjustment or suffer the consequences. Of course, no such adjustment of the sun is possible; but, at the beginning of April, it speeds up as compared to our watches, slows down again during the month of June, only to speed up once more around the first of August.

Let us suppose that we have a watch which ticks off exact mean solar seconds and that on December 24 we set it by the sun. We shall call the time defined by our watch MEAN SOLAR TIME, or simply "mean time." On December 24 the watch will agree essentially with the sun; and, if we compare it carefully with a sundial, we shall find that apparent and mean time are also equal on or about April 13, June 14, and September 1. At other times of the year, however, the two systems of time may differ by as much as 16 min. The difference—apparent *minus* mean time—is called the EQUATION OF TIME. It could, of course, be expressed by an equation, but we will simply tabulate its value for every third day of the year (Table 1). In a 3-day interval, it will be noted, the change at most is 2 min.; and for days not specified, the nearest entry will be close enough for our purposes. A plus sign denotes that the apparent time is greater than the mean, that is, the sun is fast as compared with our watch; a minus sign indicates that the sun is slow. For example, on November 1 the sun is fast by 16 min., while in February it is 13 or 14 min. slow relative to mean time.

Now, in regard to the selection of December 24 for setting our watches, let us state clearly that this date is not always the one chosen, for there is sometimes a variation of a day or so because of the insertion of February 29. The particular day selected for each year is so chosen that the sum of the negative values of the Equation of Time cancel the positive, when the sum is taken for the entire year. In discussing mean solar time, it is convenient to imagine the sun we know so well replaced by another, called the FICTITIOUS SUN,

Day of Month	Jan.	Feb.	Mar.	Apr.	May	June	July	Aug.	Sept.	Oct.	Nov.	Dec.
1	− 4	−14	−13	− 4	+ 3	+ 2	− 3	− 6	0	+10	+16	+11
4	− 5	−14	−12	− 3	+ 3	+ 2	− 4	− 6	+ 1	+11	+16	+10
7	− 6	−14	−11	− 2	+ 3	+ 2	− 5	− 6	+ 2	+12	+16	+ 9
10	− 8	−14	−10	− 1	+ 4	+ 1	− 5	− 5	+ 3	+13	+16	+ 7
13	− 9	−14	−10	− 1	+ 4	0	− 6	− 5	+ 4	+14	+16	+ 6
16	−10	−14	− 9	0	+ 4	0	− 6	− 4	+ 5	+14	+15	+ 4
19	−11	−14	− 8	+ 1	+ 4	− 1	− 6	− 4	+ 6	+15	+15	+ 3
22	−12	−14	− 7	+ 1	+ 4	− 2	− 6	− 3	+ 7	+15	+14	+ 2
25	−12	−13	− 6	+ 2	+ 3	− 2	− 6	− 2	+ 8	+16	+13	0
28	−13	−13	− 5	+ 2	+ 3	− 3	− 6	− 1	+ 9	+16	+12	− 2

TABLE 1
THE EQUATION OF TIME IN MINUTES
OR
APPARENT TIME *minus* MEAN TIME

which moves *uniformly* among the stars along the equator; its angular distance from the meridian, or its hour angle, gives us mean solar time. In other words, the fictitious sun is what our actual sun would be, if we had it repaired by a conscientious watchmaker.

THE mean time at any point on the earth is called LOCAL TIME. Since all points on the same meridian of the earth have the same meridian in the sky, it follows that such localities will have identical local times. Local time, however, varies with longitude, for the fictitious sun must cross meridians east of us before it crosses ours. We note that the fictitious sun appears to circle the globe in 24 hr., and therefore travels 1° in longitude in 4 min. Hence longitudes 1° east of us must be 4 min. ahead of us in local time; those 1° west must be 4 min. behind us. Let us reflect on the relative merits of local time. A journey eastward would provide an excellent alibi for tardiness; a journey westward would be spent setting our watches back about 4 sec. for every mile traversed.

Nevertheless, there is one real advantage of this dependence of local time on longitude, namely, the determination of longitude itself. The rotation of the earth causes the fictitious sun to pass in 1 hr. from one meridian to another 15° west in longitude; thus a difference of 1 hr. in local time implies a difference of 15° in longitude. If we know our own local time and that of Greenwich, by taking the difference in hours and multiplying by 15° we have our longitude east or west of Greenwich. Greenwich local time is obtainable by radio signals from that point. Our local time can be obtained from the sun, for example, with a sundial, changing the apparent time to mean or local time by means of the Equation of Time.

DESPITE this advantage, it would be difficult keeping appointments if our time depended upon our longitude. To eliminate part of the difficulty, in 1883 the railroads of the United States decided to employ only the local time of the meridians with longitudes a multiple of 15°, namely, 75°, 90°, 105°, and 120° west of Greenwich. The country was divided into standard time zones through which passed one of those standard meridians. All points in the same standard time zone had the same standard time. Chicago, longitude 87°.5, has for standard time the local time of the ninetieth meridian; and consequently its standard time is about 2.5×4 min. or 10 min. behind the local or mean solar time. This scheme was more universally adopted by the International Meridian Congress of 1884. The meridian of Greenwich, England, was taken as a prime meridian, and twenty-four standard meridians were established all over the globe, each differing by 15° of longitude. In America, the standard times are:

Atlantic, Colonial, or Intercolonial (Provincial), used in Nova Scotia and New Brunswick	4 hrs. slower than Greenwich
Eastern	5 hrs. slower than Greenwich
Central	6 hrs. slower than Greenwich
Mountain	7 hrs. slower than Greenwich
Pacific	8 hrs. slower than Greenwich

TIME

In a manner similar to that used for Chicago we can find the difference between local and standard times for other cities in the United States. To change local time to standard time, we add the corrections given in Table 2. If the correction is positive, it is to be added to the local time in order to obtain standard time; if negative, it is to be subtracted. It is interesting to note in passing that, although Cincinnati and Detroit are closer to the ninetieth meridian than the seventy-fifth, nevertheless they both employ Eastern Standard Time. At present, the extreme western portion of Ohio and practically all of Michigan, with the exception of Detroit, are on Central Standard Time, though in 1932 the state of Michigan petitioned for a change to Eastern Time and was refused.

For the benefit of radio enthusiasts, we list a few foreign standards of time, comparing them with both Greenwich Civil Time and Central Standard Time. Thus, at noon in Chicago, that is, 12:00 M., C.S.T., it is 6:00 P.M. in London and Paris, 7:00 P.M. in Berlin, and 3:00 A.M. in Tokyo.

The astronomer goes one step farther in denoting time. In order to avoid any possible source of error, he states the time of his observation in terms of the mean time at Greenwich—

City	Standard Time	Correction in Minutes
Baltimore, Md.	Eastern	+ 6
Boston, Mass.	Eastern	−16
Buffalo, N.Y.	Eastern	+15
Chicago, Ill.	Central	−10
Cincinnati, Ohio	Eastern	+38
Denver, Col.	Mountain	− 4
Detroit, Mich.	Eastern	+33
Flagstaff, Ariz.	Mountain	+27
Kansas City, Mo.	Central	+18
Los Angeles, Cal.	Pacific	− 7
Milwaukee, Wis.	Central	− 8
Minneapolis, Minn.	Central	+13
New Orleans, La.	Central	0
New York, N.Y.	Eastern	− 4
Philadelphia, Pa.	Eastern	+ 1
Pittsburgh, Pa.	Eastern	+20
Portland, Ore.	Pacific	+11
St. Louis, Mo.	Central	+ 1
San Francisco, Calif.	Pacific	+10
Washington, D.C.	Eastern	+ 8

TABLE 2
TABLE OF CORRECTIONS:
STANDARD−LOCAL

	Greenwich Civil Time	Central Standard Time
Spain, France	0 hr. fast	6 hr. fast
Mid-Europe	1 hr. fast	7 hr. fast
Cape Colony	1 hr. 30 min. fast	7 hr. 30 min. fast
Eastern Europe	2 hr. fast	8 hr. fast
India	5 hr. 30 min. fast	11 hr. 30 min. fast
West Australia	8 hr. fast	10 hr. slow
Japan	9 hr. fast	9 hr. slow
South Australia	9 hr. 30 min. fast	8 hr. 30 min. slow
Tasmania, East Australia	10 hr. fast	8 hr. slow
New Zealand	11 hr. 30 min. fast	6 hr. 30 min. slow

GREENWICH CIVIL TIME (or GREENWICH MEAN TIME)—which is measured from midnight, the hours being numbered from 0 to 24; e.g., 2:00 P.M., C.S.T. corresponds to 8:00 P.M. in London or 20:00 hr. G.C.T. In the past, the astronomer began his day, the *astronomical day*, at noon 12 hr. after the *civil day*, but this procedure is becoming obsolete.

To our list of time systems, we could add another, namely, DAYLIGHT SAVING, which is 1 hr. faster than standard time; but let us limit ourselves to only four modes of reckoning time: sidereal, apparent, mean (or local), and standard. To illustrate the relation between the last three systems, let us consider the following practical problem: On November 21 in Chicago we place an upright stake in the ground; and, when the shadow is due north, we decide to go home for luncheon. The problem is—at what time did we leave? The answer is noon apparent time; but, if standard time is desired (i.e. the conventional time used), we proceed as follows:

Apparent time................................... 12:00 M.
Subtract "Equation of Time," November 22.......... −:14

Mean or local time............................ 11:46 A.M.
Add correction "standard-local" for Chicago.......... −:10

Central Standard Time....................... 11:36 A.M.

A study of this problem will show the reader how, by means of a sundial, he can dispense with his watch—on a clear day.

These interrelations may be summarized as follows:

  a) Apparent time = local time + Equation of Time;
  b) Local time   = apparent time − Equation of Time.

For longitudes $x°$ west of standard meridian,

  c) Local time   = standard time − $4x$ min.;
  d) Standard time = local time + $4x$ min.

For longitudes $x°$ east of standard meridian,

  e) Local time   = standard time + $4x$ min.;
  f) Standard time = local time − $4x$ min.

TIME

As an application of these formulae consider the problem: Lost on January 13, we note that sun is due south at 11:29 A.M. C.S.T. Required, our longitude.

Solution: By Table 1, Equation of Time January 13 is -9 min. Since sun is due south, apparent time is 12:00 M. Hence, by formula (b),

$$\text{Local time} = 12:00 - (-9)$$
$$= 12:09 \text{ P.M.}$$

Our local time is thus ahead of our standard, and therefore we are east of the standard meridian. If x is the number of degrees we are east of the ninetieth meridian, then by formula (e)

$$12:09 = 11:29 + 4x,$$
$$\text{or} \quad 4x = 40,$$
$$x = 10°.$$

Our longitude is therefore $90° - 10°$, or $80°$ west of Greenwich.

There remains the problem of converting sidereal time to local time. We have already derived the rule:

$$\frac{\text{Right ascension}}{\text{of sun in hours}} = \frac{2 \times \text{number of months}}{\text{after March 21.}}$$

This rule is only approximately correct, for our sun does not move at a uniform rate eastward among the stars. We therefore replace it with the fictitious sun and write with greater accuracy:

$$\frac{\text{Right ascension of}}{\text{fictitious sun in hours}} = \frac{2 \times \text{number of}}{\text{months after the vernal equinox.}}$$

At noon local time the fictitious sun is on the meridian; and, since right ascension gives the sidereal time of *meridional transit*, it follows that

$$\frac{\text{Sidereal time at}}{\text{noon local time}} = \frac{2 \times \text{number of months}}{\text{after vernal equinox.}}$$

Thus, on June 21 when our local time clock reads 12:00 M., our star clock would indicate 6:00 hr. As the hours of mean solar time pass, so would those of sidereal time, but at a slightly faster rate, gaining 4 min. each 24 hr. Let us disregard this small gain of 10 sec. per hr. Then on June 21 at 1:00 P.M., local time, the sidereal time would be 7:00 hr.; at 2:00 P.M., local, it would be 8:00 sidereal; and so on, or in general

> Sidereal time = 2 × number of months + number of hours past
> after March 21   noon (local time).

If we obtain a result that equals or exceeds 24 hr., we subtract 24 hr.; e.g., on December 21 at 9:00 P.M. local time, sidereal time is $2 \times 9 + 9 = 27 - 24$, or 3:00 hr. Similarly, at 6:00 P.M. on the same date, sidereal time is 0:00 hr. Tables are available, such as those given in the *Nautical Almanac*, for accurate conversion of one system of time to another; but the foregoing formulae and tables will be good enough approximations for our purpose.

WE HAVE seen how the sun may be used to determine local and standard times; but, because of its size and brilliancy, observations are difficult. Then, again, it crosses the meridian only once each day. Stars, however, are mere points of light; and, for the purpose of measuring sidereal time, the transits of many different stars may be accurately observed in a single night. Both the observation of star time and its conversion into standard time for our service are made almost exclusively at national observatories—the Naval Observatory at Washington, D.C.; the Royal Observatory at Greenwich, England; L'Observatoire de Paris at Paris, France; L'Observatoire Royal de Belgique at Bruxelles, Belgium; etc. At these observatories the transits of stars are used to correct errors in man-made clocks—errors that are less than 1 part in a 100,000,000. In the Naval Observatory at Washington there is a master-clock, which will not lose or gain 1 sec. in 5 yr. and which regulates all official clocks and correlates the time signals broadcast over government radio stations. The wireless method has been perfected to such an extent that, no matter where we might be on the surface of the earth, the time signals give us Greenwich time to one-thirtieth of a second.

TIME 41

AND now, after digesting the various methods of calibrating time, we believe the reader is fully prepared for a trip around the globe. If his appetite is good, he ought to go eastward instead of west for, traveling east, one is continually setting one's watch ahead, so that meal times occur sooner. Thus, by going east around the globe, one would have not only an extra dinner but a whole day extra, as compared with those unfortunates who stay home, the converse being true for the westward journey. Somewhere, therefore, in the eastward journey we must go back a day in order that our date of homecoming will agree with that of the stay-at-homes. This is done at the INTERNATIONAL DATE LINE, an arbitrary line passing from north pole to south pole and situated, for the most part, along longitude 180°—a rather fortunate selection, for it does not pass through any large areas of land. When we cross the date line eastward, Sunday becomes Saturday and today is yesterday; going westward, we skip a day, Saturday is called Sunday and today is tomorrow. This confusion was admirably dealt with by Bret Harte in his poem "The Lost Galleon."

Let us watch a day develop on the earth. It is Friday, February 22, 5:00 A.M., C.S.T., in Chicago; it is also Friday in New York, London, and New Zealand (Long. 172°30′ E.), but the standard times for these localities are, respectively, 6:00 A.M., 11:00 A.M., and 10:30 P.M.; at longitude 180°, the international date line, it is 11:00 P.M.—Friday on the western side, Thursday on the eastern. At 6:00 A.M., C.S.T., Saturday appears at longitude 180° the instant Thursday disappears; and, half an hour later 6:30 A.M., C.S.T., Saturday reaches Wellington, N.Z. As the hours pass, Saturday occupies more and more of the globe, crowding Friday right off the earth; and at Friday 6:00 P.M., C.S.T., Saturday has attained London and covers essentially half of the globe. It reaches Chicago at midnight; and on Saturday, February 23, at 5:00 A.M., C.S.T., all but one twenty-fourth of the earth is recording the date as February 23. At 6:00 A.M., C.S.T., Saturday has circled the globe, Friday is gone, and it is again midnight at 180°, the international date line. Sunday now makes its appearance and pursues Saturday in precisely the same way as Saturday did Friday. We might therefore say that, at six o'clock this morning in Chicago, yesterday died and tomorrow was born. This statement is not strictly correct if local time is employed, for the date line does not always follow the one hundred and eightieth meridian but detours to both east and west to prevent its crossing certain islands. On the basis of local time, at 6:00 A.M., C.S.T., yesterday is still alive though tomorrow has already made its début on certain Pacific islands.

WE HAVE already tacitly assumed the reader to be familiar with our method for designating different mean solar days—the calendar. In our modern highly specialized life, most of us take the calendar for granted or regard it as a rather clumsy means for tabulating days, that is, merely a scheme for determining when we shall pay rent, taxes, and insurance or celebrate birthdays and public holidays. Not so in ancient times, for, to appreciate fully our present-day calendar, we must delve back into antiquity and examine its evolution. It is rather difficult to imagine ourselves as living in the dawn of civilization when man's prime occupation was agriculture. The most propitious time to seed was the question that faced the pioneer scientist. Without fear of sudden annihilation by speeding cars or other infernal machines, man could then look for the answer in the sky.

Ancient agricultural races noted that the sun did not always rise at the same point on the horizon but that its rising point moved north and south from the point due east. They soon discovered that this motion of the sun was intimately related to climatic conditions; for the Egyptians, the sun's rising at its most northerly point was the sign that the Nile would soon overflow its banks and irrigate their fields. Long corridors in their temples pointed at this position, so that, at daybreak on the summer solstice, and *only at that time*, would the first rays penetrate to the altar. Thus evolved not only the deification of the sun but also the construction of astronomical observatories. The priests counted the number of solar days from one summer solstice to the next and found it to be a little more than 365 days. We call this interval the TROPICAL YEAR, since it is also the period required for the sun to move from one tropic round again to

TIME

the same tropic. Present-day observations and calculations give to the tropical year, or average length of the interval between successive spring equinoxes, the value 365 days 5 hr. 48 min. 46.08 sec.; but the ancient astronomer, by watching the varying length of a vertical column or gnomon, concluded many centuries before the time of the Roman Empire that the tropical day was equal to 365.25 days! Though we no longer regard the Egyptian obelisks as precise astronomical instruments, nevertheless they enabled man to measure the year with an error of less than 1 part in 40,000.

IT IS to be regretted that our intellectual ancestors were not content with a simple division of the 365 or 366 days in the year. The moon, with its changing appearance, distracted them from the vital problem of setting agricultural dates. Religious festivals were also founded on the phases of the moon; and consequently many of the ancient calendars, being under the control of the priesthood, were predominantly lunar. The fundamental interval was the average length of time from full moon to full moon—the lunar month, or 29.530588 mean solar days. Now in a tropical year there are $365.24220 \div 29.530588 = 12.3683$ lunar months, and the astronomer-priests placated the farmers by constructing a year of twelve lunar months. Even today, such a lunar calendar is used by many of the Mohammedan nations. Since twelve lunar months equal $12 \times 29.530588 = 354.37$ days, this year contains either 354 or 355 days and is divided into twelve months, six of which contain 30 days and six 29; or seven with 30 and five with 29. With such a year, religious dates fall approximately on the same phase of the moon but in decidedly different seasons as the years pass. Since this lunar calendar differs from the tropical year by about 11 days, Mohammedans age more rapidly in their years, at the rate of 34 to our 33 yr.

Of particular interest to us is the development of the Roman calendar; in fact, the word *calendae* or *kalends*, from which "calendar" is derived, signified the first day of the Roman month. Even the names of our months are derivations of the Roman *menses*. The original calendar of the Republic commenced on the kalends of

March, the names of the months in the old order and the number of days therein being as given in the accompanying table. The sum

	No. of Days	English Equivalents		No. of Days	English Equivalents
Martius	31	March	*September*	29	September
Aprilis	29	April	*October*	31	October
Maius	31	May	*November*	29	November
Iunius	29	June	*December*	29	December
Quintilis	31	July	*Ianuarius*	29	January
Sextilis	29	August	*Februarius*	28	February

(It will be noted that the fifth and sixth months, *Quintilis* and *Sextilis*, were later dedicated to Julius Caesar and Caesar Augustus, respectively.)

total is 355 days, essentially the lunar year. Following the example of the Greeks, the Romans compensated for the deficiency in the lunar year so as to reconcile it with the seasons. Their method of intercalation was this: Every second and fourth year they were supposed to insert 22 or 23 days in the latter portion of February, which made the average year about 1 day greater than the tropical year. In the course of time, this system also got out of harmony with the seasons; adjustments were made now and then by the pontifices, but, through ignorance or negligence on their part, the calendar was thrown into confusion before the last century of the Republic.

IN THE year 46 B.C. Julius Caesar sought the advice of the Alexandrian astronomer Sosigenes as to the reformation of the calendar. The result was the JULIAN CALENDAR, which is essentially the same as the one we tear monthly. Three out of four years he made 365 days in length while the fourth, or leap year, was 366 days, as it now is. The average, or Julian year, is thus 365.25 days, which was the length then assigned to the tropical year. Julius Caesar avoided the old scheme of intercalation by increasing the length of certain months. To January, August (*Sextilis*), and December, 2 days were added; and to April, June, September, and November, 1 day in ordinary years, while in leap years February had 29 days.

TIME

For our peculiar division of the year into twelve months we may thank both the moon and Caesar, for he evidently did not have the courage to discard the twelve months, as he merely added a day or two to the old lunar months, removing them from the moon's control, so that they totaled 365 or 366 days. About a century prior to this reformation of the calendar, the Romans had changed the beginning of their official year to January 1, though the kalends of March was still regarded as the commencement of the religious year, and Julius Caesar did not alter this practice. He selected as January 1 the day of the new moon immediately following the winter solstice, which, in 46 B.C., came a little more than a week after the winter solstice—a position it still occupies.

IT FOLLOWS from our observations and calculations of the tropical year that the Julian calendar is too long by $365.25 - 365.24220 = .0078$ days, so that after four centuries the seasons would occur $400 \times .0078 = 3.12$ days earlier. Despite the fact that it takes many centuries to alter appreciably the positions of the seasons in the Julian calendar, in 1582 Pope Gregory XIII attempted to correct this slight discrepancy. Since the error in the Julian year amounts to a fraction more than 3 days in four centuries, Pope Gregory modified the Julian calendar by arranging that the closing years of the centuries 1600, 1700, etc., should not be leap years unless the number of the century were divisible by 4; that is to say, the years 1600, 2000, 2400, etc., were to be leap years as in the Julian calendar but 1700, 1800, 1900, 2100, etc., were to be common years of 365 mean solar days.

The change to the GREGORIAN CALENDAR probably was not made for agricultural requirements but for religious purposes, for, in the course of centuries, the Julian had the effect of making the equinox fall at a time of the year different from that at which it had been arranged to fall by the festivals of the Christian church. Pope Gregory assumed that the correct arrangement was that made by the Council of Nicaea in 325 A.D. In the interval 325 A.D. to 1582 A.D., that is, 1,257 years, the displacement of the religious dates relative to the seasons amounted to $1,257 \times .0078 = 9.8$ days; and, to make the arrangement of religious dates correspond with those in 325 A.D., Gregory XIII decreed that 10 days should be omitted in the year 1582. This was accomplished by calling the day following October 4 the fifteenth day of October.

The alteration, however, was not at first universally accepted, and it was not until 1752 that England and her colonies began to use the Gregorian calendar. Since the year 1700 was a leap year in the Julian calendar but not in the Gregorian, a total of 10 plus 1, or 11, days were omitted in accomplishing the change, by having September 14 follow September 2. Prior to 1752, the year in England began on March 25; but from then on, January 1 has been New Year's Day. We in the United States celebrate Washington's birthday on February 22 (Gregorian, or New Style), though he was born February 11, 1732 (Julian, or Old Style). Practically all the nations of the world have been forced to adopt the clumsy Gregorian calendar. Japan accepted it in 1873, the Chinese Republic in 1912, Turkey in 1917, Roumania in 1919, and Greece and the Greek Orthodox church in 1923.

THE question naturally arises whether or not we should make any further changes in our present year—the Gregorian year. In 400 years there are 100−3 leap years, making the average length of the Gregorian year 365.2425 mean solar days, or $365^d5^h49^m12^s$ of mean solar time, which differs from the tropical year ($365^d5^h48^m46^s.08$) by less than half a minute. Of course, in 3,000 years this will amount to a day; and so we find the Union of Socialist Soviet Republics, although waiting until 1918 to discard the Julian calendar, adopting a leap-year rule for centuries even more accurate than the Gregorian. The error of their calendar is only about 3 sec. in a year—1 day in 30,000 yr.! The rule is that century years shall be leap years provided their numbers, when divided by 9, give a remainder of 2 or 6. The Eastern and Gregorian calendars, however, will agree until 2800 A.D.

We may well question the necessity for such refinements. When we reflect that in England the change from Julian to Gregorian, despite careful and just legislation, led to rioting and bloodshed in

TIME

many cities, with the mob crying "Give us back our eleven days," we wonder whether we should attempt to make such changes, merely for the sake of mathematical perfection. Some worthy Protestants, we are told, were so scandalized by the adoption of the Gregorian calendar that they continued to observe Good Friday according to Julian, or Old Style, reckoning. Barred from the church, they were compelled to read the services at home. Nor were they alone in their revolt, for it was maintained by others of the same school of thought, that on the morning of December 25, 1752 (O.S.), even the cattle fell on their knees!

OBVIOUSLY, in computing time intervals involving Old Style and New Style, considerable confusion is likely to arise. This may be avoided by the use of the JULIAN DAY NUMBER, computed by Joseph Scaliger in 1582 and so named, not for Julius Caesar, but in honor of its inventor's father. This device employs cycles of 7,980 years, the Julian period, in place of single years, the last cycle beginning January 1, 4713 B.C. (Scaliger determined the Julian period by making it a least common multiple of three cycles used by the Romans; the beginning of the last Julian period he tried to fix at a date upon which all three cycles began together.) The Julian day number is the number of days which have passed since January 1, 4713 B.C., the Julian day number for January 1, 1935 A.D., being 2,427,804. These day numbers are very convenient for setting historical events. Again, by simple subtraction, the interval in days between two eclipses can be found if we know their Julian day numbers. Although our civil day begins at midnight, the Julian day, by international agreement, commences at noon.

DESPITE the fact that non-Christian peoples are rapidly adopting the Gregorian as their national calendar, nevertheless they still employ other calendars for ecclesiastical purposes. Some of them, such as the Jewish and Mohammedan calendars, are based on the lunar month; others on the Julian year, and the years are numbered from some historical or legendary event. The Mohammedan Era begins with the Hegira, or flight of Mohammed from

Mecca, 622 A.D.; the Jews reckon from 3761 B.C., their traditional date for the Creation; the ancient Romans, from the legendary date of the founding of Rome, 753 B.C. The Christian Era begins with the date of the birth of Christ, as determined in the sixth century by Dyonisius Exiguus, a monk of Scythia; but, according to calculations based on the computed date of an eclipse, Jesus Christ was born in the year 4 B.C.

In *The American Ephemeris and Nautical Almanac* for 1935 we find, under "Chronological Eras," that

The year 1935, which comprises the latter part of the 159th and the beginning of the 160th year of the independence of the United States of America, corresponds to—
The year 6648 of the Julian period.
" 7444 of the Byzantine era, the year commencing on September 1.
" 5696 of the Jewish era, the year commencing on September 28 or, more exactly, at sunset on September 27.
" 2688 since the foundation of Rome, according to VARRO.
" 2684 since the beginning of the era of Nabonassar, which has been assigned to Wednesday, the 26th of February of the 3967th year of the Julian period; corresponding in the notation of chronologists to the 747th and, in the notation of astronomers, to the 746th year before the birth of Christ.
" 2711 of the Olympiads, or the third year of the 678th Olympiad, commencing in July, 1935, if we fix the era of the Olympiads at $775\frac{1}{2}$ years before Christ, or near the beginning of July of the year 3938 of the Julian period. (*Note:* The Olympiad was the interval of four years between the Olympic Games and was used by the ancient Greeks to reckon years until A.D. 394.)
" 2247 of the Grecian era, or the era of the Seleucidae, which began near the vernal equinox of the year, $-311 = $ B.C. $312 = 4402$ of the Julian period. (*Note:* In present day usage of the Syrians the year begins on September 1, 1935.)
" 1651 of the era of Diocletian.
" 2595 of the Japanese era and to the 10th year of the period entitled Showa.
The year 1354 of the Mohammedan era, or the era of the Hegira, begins the 4th day of April, 1935.

Obviously, the query, "What year is it?" can be answered in a multitude of ways.

In both Protestant and Catholic countries, an ecclesiastical calendar, a lunisolar calendar, is used for regulating the dates of church feasts and observances. Our present Easter Sunday and the religious dates associated with it are good illustrations of lunar religious festivals. Astronomically, the rule for fixing the date of Easter is: *The first Sunday after the first full moon immediately following the vernal equinox.* Since our present calendar disregards the lunar phases, the date of the occurrence of this full moon varies erratically from year to year. There are 12.3683 lunar months or full moons per tropical year; and in 19 yr. there will be $12.3683 \times 19 = 234.9977$, or almost exactly 235, full moons. If we have a full moon on January 1 of a particular year, then 19 yr. later we will again have a full moon within a day (to allow for leap years) of January 1; in other words, every 19 yr. full moon will occur at approximately the same calendar dates. This interval, known as the METONIC CYCLE, was discovered by Meton in 433 B.C. and is fundamental in the ecclesiastical calendar. The GOLDEN NUMBER, so named because the plan of Meton was inscribed on public monuments in letters of gold, is 1 plus the remainder obtained by dividing the year number by 19; thus the Golden Number for 1935 is 17. All years having the same Golden Number will have full moons on approximately the same dates for many centuries to come.

THERE remains the division of the year into weeks. The origin of the 7-day week is not known with certainty, but the names of the days are almost definitely of astronomic origin. The ancient Chaldeans noticed that seven heavenly bodies (the sun, moon, and five planets) moved among the stars; and they supposed their distance from the earth decreased in the order: Saturn, Jupiter, Mars, Sun, Venus, Mercury, and the Moon. Each hour of the day was associated with one of those objects in the order given, with repetitions of course; and the day was identified by the name of its first hour. Thus, the first hour of Saturday, as well as the day itself, was named after Saturn; the second hour of Saturday was named after Jupiter; the third, after Mars; and so on until the eighth hour, which again corresponded to Saturn; and the process repeated. In this manner, the first hour of Sunday was named after the Sun and, by continuing the process over the 7 days of the week, we find the following association:

CELESTIAL OBJECT	DAY OF THE WEEK	
	Latin	English
Sun	Dies Solis	Sunday
Moon	Dies Lunae	Monday
Mars	Dies Martis	Tuesday
Mercury	Dies Mercurii	Wednesday
Jupiter	Dies Jovis	Thursday
Venus	Dies Veneris	Friday
Saturn	Dies Saturni	Saturday

We have all noticed how our birthday selects different days of the week from year to year. In a common year there are 52 weeks plus 1 day; in a leap year, 52 weeks plus 2 days; and, consequently, to the delight of calendar-manufacturers, each successive year begins 1 or 2 days later in the week. For example, January 1 was a Friday in 1932, a Sunday in 1933, a Monday in 1934, and a Tuesday in 1935. In the case of common years, one of the first seven letters of the alphabet, the DOMINICAL LETTER, is often used to denote the date corresponding to the first Sunday in January. The dominical letter for 1933 is A; 1934, G; 1935, F, the order of the letter in the alphabet giving the date. In the case of leap years, two dominical letters are used, the first indicating Sunday till the 29th of February and the other for the rest of the year. The dominical letters for the years 1933, 1934, 1935, 1936, 1937, 1938, 1939, 1940, are respectively A, G, F, ED, C, B, A, GF—the pianist will have no difficulty in discovering the order.

For the benefit of the ultra-thrifty, we note that years with the same dominical letter or letters have identical calendars, as far as the position of the days of the week are concerned; furthermore, in the case of leap years, we may use parts of two common years. For example, in 1940, dominical letters GF, January and February could be salvaged from the calendar for 1934 (dominical letter G), and the remainder from 1935. It is not a difficult task to construct a socalled perpetual calendar in terms of dominical letters. With this

PERPETUAL CALENDAR

DOMINICAL LETTER

A	B	C	D	E	F	G							
Jan., Oct.	May	Aug.	Feb., Mar., Nov.	June	Sept., Dec.	Apr., July	Sun.	Mon.	Tues.	Wed.	Thur.	Fri.	Sat.
Apr., July	Jan., Oct.	May	Aug.	Feb., Mar., Nov.	June	Sept., Dec.	Sat.	Sun.	Mon.	Tues.	Wed.	Thur.	Fri.
Sept., Dec.	Apr., July	Jan., Oct.	May	Aug.	Feb., Mar., Nov.	June	Fri.	Sat.	Sun.	Mon.	Tues.	Wed.	Thur.
June	Sept., Dec.	Apr., July	Jan., Oct.	May	Aug.	Feb., Mar., Nov.	Thur.	Fri.	Sat.	Sun.	Mon.	Tues.	Wed.
Feb., Mar., Nov.	June	Sept., Dec.	Apr., July	Jan., Oct.	May	Aug.	Wed.	Thur.	Fri.	Sat.	Sun.	Mon.	Tues.
Aug.	Feb., Mar., Nov.	June	Sept., Dec.	Apr., July	Jan., Oct.	May	Tues.	Wed.	Thur.	Fri.	Sat.	Sun.	Mon.
May	Aug.	Feb., Mar., Nov.	June	Sept., Dec.	Apr., July	Jan., Oct.	Mon.	Tues.	Wed.	Thur.	Fri.	Sat.	Sun.
							1	2	3	4	5	6	7
							8	9	10	11	12	13	14
							15	16	17	18	19	20	21
							22	23	24	25	26	27	28
							29	30	31				

Omit 31—February, April, June, September, and November.
Omit 30—February.
Omit 29—February for common years.

perpetual calendar we may quickly ascertain the day of the week for any date in the past or future.

Suppose we wish to find the day of the week corresponding to October 28, 1937. The dominical letter is C, and we look for October under this letter. On a line with this month, we find the days of the week that correspond to the dates for October, 1937; the 28th day is a Thursday. In the case of leap years, the first dominical letter is used for January and February and the second for the remaining

DOMINICAL LETTERS (NEW STYLE)

	Century Numbers				1300 1700 2100 2500 2900	1400 1800 2200 2600 3000	1500 1900 2300 2700 3100	1600 2000 2400 2800 3200
	0				C	E	G	BA
	1	29	57	85	B	D	F	G
	2	30	58	86	A	C	E	F
	3	31	59	87	G	B	D	E
	4	32	60	88	FE	AG	CB	DC
	5	33	61	89	D	F	A	B
	6	34	62	90	C	E	G	A
	7	35	63	91	B	D	F	G
	8	36	64	92	AG	CB	ED	FE
Year	9	37	65	93	F	A	C	D
	10	38	66	94	E	G	B	C
	11	39	67	95	D	F	A	B
	12	40	68	96	CB	ED	GF	AG
Numbers	13	41	69	97	A	C	E	F
	14	42	70	98	G	B	D	E
	15	43	71	99	F	A	C	D
	16	44	72		ED	GF	BA	CB
	17	45	73		C	E	G	A
	18	46	74		B	D	F	G
	19	47	75		A	C	E	F
	20	48	76		GF	BA	DC	ED
	21	49	77		E	G	B	C
	22	50	78		D	F	A	B
	23	51	79		C	E	G	A
	24	52	80		BA	DC	FE	GF
	25	53	81		G	B	D	E
	26	54	82		F	A	C	D
	27	55	83		E	G	B	C
	28	56	84		DC	FE	AG	BC

months of the year. In this manner, we identify April 7, 1936 (dominical letters *ED*), through the letter *D* as a Tuesday. In order that the reader will be able to use this calendar for another thousand years or so, we have listed the dominical letters for the years 1300 to 3299 A.D., inclusive. If the interval is too short, he may readily extend the table by adding the desired century numbers.

Many other applications of this perpetual calendar could be given. A lady, after careful questioning, informs us that she was born in the twentieth century on a Saturday and that her birthday is June 21. By a careful study of our perpetual calendar we find that the dominical letter for her birth year is *E* if a common year, or in the case of a leap year the second letter is *E*. Taking all other factors into consideration in conjunction with our table of dominical letters, we try to guess the year of her birth, our choice prior to 1941 being limited to 1902, 1913, 1919, 1924, and 1930.

It should be noted that the preceding table is based upon the New Style, or Gregorian, calendar. The following table gives the difference between the Gregorian date and the corresponding date in the Julian, or Old Style, reckoning.

Interval (A.D.)	Number of Days To Be Added to Julian Date To Convert into Gregorian
1301–1400*	8
1401–1500*	9
1501–1700*	10
1701–1800*	11
1801–1900*	12
1901–2100*	13

The years with asterisks are those leap years in the Julian calendar that are common years in the Gregorian. To avoid confusion, express the dates for these particular years as the number of days that have elapsed since the beginning of the year. It follows that if Columbus discovered America on October 12, 1492 (O.S.), he discovered it also on October 21, 1492 (N.S.).

Reform of the Gregorian calendar has been advocated for many years both in this country and abroad, and the movement has at last gained sufficient momentum to receive the attention of the League of Nations. The Advisory and Technical Committee for Communication and Transit of the Council reported in 1931 that the Catholic and Anglican churches and most of the Protestant churches of the United States had no insurmountable objections to the proposed changes but that objections were voiced by representatives of the Jewish faith, Seventh-Day Adventists, and Seventh-Day Baptists. To rectify in some degree the inequality in the lengths of the months and to compel Sunday to fall always on the first of the year, the Rational Calendar Association of England and the World Calendar Association of America have sponsored a twelve-month calendar in which January, April, July, and October always begin on Sunday and have 31 days apiece; February, May, August, and November commence on Wednesday with 30 days each; March, June, September, and December start on Friday, also with 30 days each. The odd 365th day of the year, called "Year Day," is considered as an extra Saturday between December 30 and January 1; in leap years, an additional "Leap Year Day," also a Saturday, is inserted between June 30 and July 1.

Though the churches have been largely responsible for our present-day calendar, it seems fitting in this present generation to consult other activities before adopting a new system. Every day millions of mortals waste precious moments in mental gymnastics endeavoring to determine the proper date. If we *must* submit to a new calendar, let it be a simple one this time, such as the thirteen-month calendar. According to a report of the International Calendar Reform Association, one hundred and forty large concerns in the United States, including Sears, Roebuck and Company, Eastman Kodak Company, Jewel Tea Company, Hotel New Yorker, and Loew's Theaters, are using a thirteen-month calendar in their accounting departments. With this calendar it would not be necessary to tear off the past month on the first of the next, for the thirteen-month calendar *always* reads as indicated on the next page.

The name "Sol" has been submitted for the proposed new month, and it is suggested that it be placed between June and July; but it would be a pity to inflict on future generations our present

EVERY MONTH

Sun.	Mon.	Tues.	Wed.	Thur.	Fri.	Sat.
1	2	3	4	5	6	7
8	9	10	11	12	13	14
15	16	17	18	19	20	21
22	23	24	25	26	27	28

Year End Day, December 29; Leap Year Day, June 29

absurd nomenclature of the months. Even Roman numerals would be far superior to a system that names its first month after a two-faced god of beginnings and then terminates the year with cognomens that indicate a sad misunderstanding of the fundamentals of arithmetic.

BEFORE leaving the discussion of the calendar, it is interesting to point out that the tropical year, which is the basis for our calendar, is not exactly the length of time it takes for the earth to go around the sun. The latter period, which is also the length of time it takes the sun to complete its apparent circuit among the stars, is called the SIDEREAL YEAR and is 20^m23^s42 longer than the tropical year. The vernal equinox, which we described as a fixed point among the stars, actually moves very slowly westward; and while the sun is on its annual eastward journey, the vernal equinox comes up to meet it part way—but an extremely small part—thus shortening the tropical year. We call this motion PRECESSION OF THE EQUINOX and trace its cause to the attraction of the moon and sun on the equatorial bulge of the earth. In the next chapter we shall again encounter effects of precession.

We shall content ourselves by noting here that the difference in tropical and sidereal years is such that, in 25,800 calendar years, we make exactly 25,799 revolutions about the sun. The interval 25,800 tropical years is the length of time it takes the vernal equinox to return to its former position among the stars, and is known as the

GREAT YEAR or PLATONIC YEAR. We might also mention that precession of the equinoxes makes the sidereal day, or the interval between successive transits of the vernal equinox across the meridian, less, by one-hundredth of a second, than the period of rotation of the earth relative to the stars. Though this difference seems trivial, nevertheless precession is one of the many factors that must be considered in the accurate determination of time.

W E HAVE seen how our modes of reckoning time are an outgrowth of tradition, practicability, and the motions of the earth. It is natural to question the reliability of the last factor—the earth. Though no man-made clock has succeeded in detecting any irregularities in the earth's rotation, attempts have recently been made to check the earth-clock with another celestial-clock—the revolution of the moon about the earth. By an exhaustive study of the moon, it has been found that there are slight discrepancies between the observed position of the moon and that predicted by theory: the moon appears always to be a trifle ahead of the theoretical schedule. The problem is one of extreme complexity, and it may be that all factors have not been considered. Nevertheless, there is good reason to believe that the moon is on time and that our watch, the earth, is running slow. Tides generated by moon and sun act as a brake on the earth, but there is no cause for alarm or to agitate for redivision of the working day, for E. W. Brown calculates that not until 100,000 A.D. will the day be 1 sec. longer.

THE SKY

> And that inverted Bowl we call The Sky,
> Whereunder crawling coop't we live and die,
> Lift not thy hands to *It* for help—for It
> Rolls impotently on as Thou or I.
>   OMAR KHAYYÁM, *Rubáiyát*

THE automobile has been credited with bestowing a multitude of blessings upon mankind, one of which we shall emphasize. Through the constant and widespread use of road maps, even the veriest tyro is able to plan an itinerary sans mental strain and uncertainty. On the cartograph presented to us with the compliments of a large oil refinery certain lines are drawn to indicate latitude and longitude—a scheme for locating spots on our spherical earth that we owe to Hipparchus, the Father of Astronomy. A little more than twenty centuries ago, Hipparchus became also the Father of true Geography by applying his method of mapping the sky to the earth itself.

The dual rôle of astronomer and geographer should cause no surprise, for the problems of locating stars in the sky and positions on the surface of the earth are essentially the same. In this connection, we remember another astronomer-geographer, Claudius Ptolemy, who continued Hipparchus' work of charting both earth and sky in the second century A.D. Ptolemy's determination of the position of stars, however, was far more accurate than his catalogue of latitudes and longitudes; his latitudes were tolerably correct, but his longitudes were wide of the truth, making his estimate of the extent of the known world from east to west much too great. Nevertheless, his eight books on geography were accepted as *the* textbooks of this science from the middle of the second to the beginning of the sixteenth century, while his faulty longitudes encouraged a belief in the practicability of a western passage to the Indies. An astronomer's mistake, uncorrected for thirteen hundred and fifty years, led to the discovery of America by Columbus!

WE TURN now to the educational possibilities of aeroplane transportation. On a clear night we fly between earth and sky—above are the lights of stars; below, those of cities. We pilot our plane by man-made beacons or direct our course by the more permanent stars, the choice of method depending on whether we wish to look up or down. At all times some point on the earth is directly below a particular star in the sky and, as we fly along, we begin to understand that correspondence between earth and sky established by Hipparchus.

To simplify the problem of navigating a plane, we adopt his picture of the universe, as far as earth and stars are concerned. Though the celestial bodies differ enormously in distance from the earth, they are so incomprehensibly far away that, for our purposes, we might just as well regard them as being placed on one and the same sphere—the celestial sphere. At the center of this huge sphere we have a tiny sphere—the earth. So large is the celestial sphere, and so small our globe, that the appearance of the former would be just the same whether viewed from earth's surface or earth's core. In the more generally accepted plan of the ancients, including Hipparchus, the inner sphere (earth) was stationary while the celestial sphere turned westward, carrying the stars along with it in its motion. The pedagogical advantages of this scheme encourage us to shelve our proofs of the earth's rotation, so that, for the major por-

THE SKY

tion of this chapter, we consider ourselves as being indeed on *terra firma*.

Let us reverse the mental procedure of Hipparchus: starting with familiar circles and points on the earth, we find their counterpart in the sky. As we have previously noted, directly over the poles of the earth are the north and south celestial poles; and it is about these two points that the celestial sphere appears to rotate westward. By sighting upward along a plumb line we locate the zenith—or, rather, *our* zenith—the point in the celestial sphere that hovers always directly overhead. At the north pole of the earth, zenith and north celestial pole are identical; but, as we travel on the earth's curved surface, the north celestial pole descends relative to our zenith, until, at the equator, it is due north on the horizon, its behavior being in accordance with the previously derived rule:

Altitude (angular height above horizon) of celestial pole equals the latitude of the observer.

In the northern hemisphere the north celestial pole is above the horizon and the south celestial pole an equal angular distance below; these conditions are naturally reversed in the southern hemisphere. Joining these three points, the two celestial poles and the zenith, is the celestial meridian, that important semicircle we repeatedly encountered in time measurements. The other half of this circle is also worthy of a name, and we shall call it the LOWER MERIDIAN, for it passes through the NADIR, which is the point directly below us as opposed to the zenith, which, as we have mentioned, is the point directly above us. We in the middle-northern latitudes see that small portion of the lower meridian which lies between the north celestial pole and the north point on the horizon. In our study of time we learned that, when the sun crosses our meridian (or, more precisely, when the center of the fictitious sun transits our standard meridian), A.M. (*ante meridiem*) becomes P.M. (*post meridiem*), while its lower meridional passage marks the end of P.M. and the beginning of A.M.

The problem of locating the stars could be greatly simplified by stopping the westward turning of the celestial sphere, for then we could define the position of a star by the latitude and longitude of the terrestrial point directly under it. But we must tolerate this daily motion of the stars and study the consequences. The sky rotates around the pole, and therefore the spot on earth directly below a particular star moves along a definite parallel of latitude. It follows that, if a star is directly over the pole of the earth, it will stay there; if above the equator, it will remain above on the CELESTIAL EQUATOR at least as far as our lifetime is concerned. Thus, despite rotation, the latitude of the point on earth directly below a star remains unchanged. This latitude associated with a star we shall call DECLINATION. For example, the star Algol, which means "The Demon," has a declination of $40°.7$ and therefore passes directly over Naples in Italy and New York City, crossing the meridian at the zenith for all points on Lat. $40°.7$ N. The declinations of stars that travel directly over latitudes in the southern hemisphere are written as negative numbers; Sirius, the brightest star in the sky, has a declination $-16°.6$ and is seen crossing the meridian at the zenith in South Central Brazil, Bolivia, Southern Peru, Rhodesia, and Northern Australia.

THE maximum angular height a star attains in its journey across the sky is found from its declination and the observer's latitude. As already noted, if the declination is equal to the latitude of the observer, the star will pass through his zenith; but if the declination exceeds his latitude, then the star will be seen overhead at latitudes north of the observer and will therefore cross the meridian north of his zenith. Similarly, if the declination is less than the latitude, the star will cross to the south. The angular distance from zenith to star, or ZENITH DISTANCE, at meridional passage is for all cases the difference between the latitude of the observer and the declination of the object. This important result we reiterate in a formula:

Zenith distance at meridional transit:
   If declination exceeds latitude,
      Zenith distance=declination MINUS latitude
and star crosses meridian north of zenith.

THE SKY

> If latitude exceeds declination,
> Zenith distance = latitude MINUS declination
> and star crosses meridian south of zenith.

By a simple subtraction, therefore, we find that the star Deneb, declination 45°, at Lat. 40° N. crosses the meridian 5° north of the zenith; while, for this same latitude, Sirius, declination $-16°.6$ at its highest, is $40° - (-16°.6) = 56°.6$ south of the zenith. To convert zenith distance (angular distance from zenith) into altitude (angular distance from horizon), we note that the zenith is 90° from the horizon and hence

> Altitude = 90° *minus* zenith distance,
> Zenith distance = 90° *minus* altitude.

For example, the maximum altitude for Deneb in latitude 40° is 85°; for Sirius, 33°.4. A zenith distance greater than 90° or a negative altitude implies that the object is below the horizon.

In passing, we wish to point out that these same formulae enable us to find latitude, given the zenith distance or altitude of a star of known declination. To illustrate the method, suppose the observer watches the star Deneb, declination 45°, cross his meridian south of the point overhead at a zenith distance of 20° (altitude 70°). Direct substitution in the formula yields the equality

> 20° = latitude *minus* 45°,

and we conclude that the observer is in Lat. 65° N.

Because the stars in their daily motion are always over the same parallel of latitude, the problem of determining latitude is comparatively simple. It is not surprising, therefore, that Ptolemy and other early geographers were successful in their assignation of latitudes. But the longitude of a point directly below a star is constantly changing, so that, to measure the longitude of terrestrial points, we must either make simultaneous observations at widely separated places on the earth or take due account of the elapse of time from one observation to the next. Nowadays either procedure could be adopted, but obviously this was not the case two thousand years ago. It is no wonder that Ptolemy erred in his determination of longitudes, for he based his figures on estimated distance traversed by caravans and galleys! Had transportable timepieces, such as the pocket watch, been discovered in their day, these early geographers would not have overestimated the east and west extension of Europe and Asia; had telephone service or radio been popular a thousand years ago, Columbus might not have set out on his journey.

WE HAVE answered the question as to *where* a star will cross our meridian, and our next step is to find the *time* of its meridional passage. But we have already solved this problem in chap-

Star	Interpretation	Constellation Name	Magnitude	Right Ascension	Declination
Sirius.............	Sparkling star	α Canis Majoris	−1.6	6^h41^m	−16°6
*Canopus..........	Ancient city in Lower Egypt	α Carinae	−0.9	6 22	−52.7
*Rigil Kentaurus.....	Foot of the Centaur	α Centauri	−0.2	14 34	−60.5
Vega..............	Falling	α Lyrae	0.1	18 34	38.7
Capella...........	Little she-goat	α Aurigae	0.2	5 11	45.9
Arcturus..........	Bear-keeper	α Boötis	0.2	14 12	19.6
Rigel.............	Foot (of the giant)	β Orionis	0.3	5 11	− 8.3
Procyon...........	Before the dog	α Canis Minoris	0.5	7 35	5.4
*Achernar.........	End of the river	α Eridani	0.5	1 35	−57.7
*Agena............	?	β Centauri	0.9	13 58	−60.0
Altair.............	Flying eagle	α Aquilae	0.9	19 47	8.6
Betelgeuse........	Shoulder of central one	α Orionis	0.9	5 51	7.4
*Acrux............	Alpha of the (Southern) Cross	α Crucis	1.0	12 22	−62.6
Aldebaran.........	Following (the Pleiades)	α Tauri	1.1	4 31	16.3
Spica.............	Ear of grain	α Virginis	1.2	13 21	−10.7
Pollux............	Boxer	β Geminorum	1.2	7 40	28.2
Antares...........	Like Ares (Mars)	α Scorpii	1.2	16 24	−26.2
Fomalhaut........	Fish's mouth	α Piscis Australis	1.3	22 53	−30.1
Deneb............	Hen's tail	α Cygni	1.3	20 39	45.0
Regulus...........	Little king	α Leonis	1.3	10 4	12.4

* Not visible in Lat. 36°–90° N.

TABLE 3

THE FIRST-MAGNITUDE STARS IN ORDER OF BRIGHTNESS

ter 2, for the right ascension of a star is the same as the sidereal time of its transit across the meridian. In Table 3 are given the right ascensions and other data for the twenty brightest stars in the sky.

The column headed "Right Ascension" is a time-table for meridional passage in terms of sidereal time. This stellar schedule informs us that Sirius is due on our meridian at 6^h41^m sidereal time;

THE SKY

Canopus, at 6^h22^m sidereal time; and so on down the list. But most of us set our watches by standard time in accordance with general usage, and thereby we complicate to some extent the problem of identifying the stars. In order that the student be spared the expense of maintaining a sidereal clock, we reiterate the rule for obtaining sidereal time:

Sidereal time in hours is equal to twice the number of months past the vernal equinox PLUS the number of hours past noon (less 24 if the result equals or exceeds 24).

By "noon" we mean *noon local time;* and, since our watches are adjusted for standard time (or Daylight Saving Time, which is 1 hr. ahead of standard), for precise results we must first convert standard time into local in the manner outlined in the previous chapter (Table 2). However, inasmuch as we are only interested at this point in the approximate position of celestial objects for the purpose of their identification, we shall simplify our computations by taking as our noon that which is defined by our watch—12:00 M., Standard Time, 1:00 P.M. Daylight Saving Time. We therefore consider just two clocks—our watch and the stars. We know that star time runs a trifle fast as compared to clock time; but, to find its value, we merely double the number of months and fraction thereof past March 21 and add on the number of hours past noon as indicated by our watch. This figure, "twice the number of months after the vernal equinox," may be obtained by means of a little mental arithmetic or read from Table 4.

Obviously, the figures listed can only be approximate, but they will be found close enough for our purpose. In Table 4 the first number indicates hours and the second minutes; e.g., on January 27 we add 20^h31^m to the time that has elapsed since noon, so that, at 3:10 P.M. on that date, sidereal time is 23^h41^m. To convert sidereal time to "hours and minutes past noon," the table entry under the particular day in question is subtracted from the given sidereal time. In computing, it should be remembered that sidereal time is numbered from 0 to 24, so that 24 hr. sidereal is equivalent to 0 hr., 25 hr. sidereal to 1 hr. and so on. In converting sidereal time to hours past noon, it may happen that the amount to be subtracted as given in the table exceeds the specified sidereal time. In this case, 24 hr. are added to the sidereal time *before* performing the indicated subtraction; e.g., when the sidereal time on February 11 is 10 hr. the time that has elapsed since the previous noon is 10 *plus* 24 *minus* $21:30 = 12^h30^m$, i.e., standard time is approximately 12:30 A.M. There is a variation of a few minutes in the tabular entries due to

Day	Jan.	Feb.	Mar.	Apr.	May	June	July	Aug.	Sept.	Oct.	Nov.	Dec.
1......	18:48	20:50	22:41	0:43	2:41	4:44	6:42	8:44	10:46	12:45	14:47	16:45
3......	18:56	20:58	22:49	0:51	2:49	4:51	6:50	8:52	10:54	12:52	14:55	16:53
5......	19:04	21:06	22:57	0:58	2:57	4:59	6:58	9:00	11:02	13:00	15:03	17:01
7......	19:12	21:14	23:05	1:06	3:05	5:07	7:06	9:08	11:10	13:08	15:10	17:09
9......	19:20	21:22	23:12	1:14	3:13	5:15	7:13	9:16	11:18	13:16	15:18	17:17
11......	19:28	21:30	23:20	1:22	3:21	5:23	7:21	9:24	11:26	13:24	15:26	17:25
13......	19:36	21:38	23:28	1:30	3:29	5:31	7:29	9:31	11:34	13:32	15:34	17:32
15......	19:43	21:46	23:36	1:38	3:37	5:39	7:37	9:39	11:42	13:40	15:42	17:40
17......	19:51	21:54	23:44	1:46	3:44	5:47	7:45	9:47	11:49	13:48	15:50	17:48
19......	19:59	22:01	23:52	1:54	3:52	5:55	7:53	9:55	11:57	13:56	15:58	17:56
21......	20:07	22:09	0:00	2:02	4:00	6:02	8:01	10:03	12:05	14:03	16:06	18:04
23......	20:15	22:17	0:08	2:10	4:08	6:10	8:09	10:11	12:13	14:11	16:14	18:12
25......	20:23	22:25	0:15	2:18	4:16	6:18	8:16	10:19	12:21	14:19	16:21	18:20
27......	20:31	22:33	0:23	2:26	4:24	6:26	8:24	10:27	12:29	14:27	16:29	18:28
29......	20:39	......	0:31	2:33	4:32	6:34	8:32	10:34	12:37	14:35	16:37	18:36
31......	20:47	......	0:39	......	4:40	......	8:40	10:42	......	14:43	......	18:44

TABLE 4

APPROXIMATE SIDEREAL TIME AT NOON—AMOUNT TO BE ADDED TO HOURS
PAST NOON TO OBTAIN SIDEREAL TIME

the shift in the date of the vernal equinox, but this deviation need not concern us.

Given the right ascension of the object, with the aid of the table and by means of a simple subtraction we can determine the approximate standard time of its meridional passage. Suppose that the date is May 6; Table 4 indicates that the sidereal time at noon is about 3:00; hence at 6:00 P.M. on this date it is 9:00 sidereal time; at midnight, 15:00. On consulting the stellar time-table, we conclude that Arcturus, Spica, as well as all other stars with right ascension between 9:00 and 15:00, will cross our meridian on May 6 between 6:00 P.M. and midnight; to find the time of their meridional passage on the date in question, we merely subtract 3:00 from the given right ascension of the stars.

THE SKY

One method of identifying stars is to locate them at the instant of their meridional passage. A neophyte in astronomy, residing in Lat. 40° N., might construct, in anticipation of a clear night on May 6, the accompanying table from the rules here given. Weather permitting, our observer places himself outdoors in a comfortable reclining chair facing the south and patiently awaits the passage of these stars across the meridian, thereby checking his computations and familiarizing himself with the objects.

Star	Local Time of Transit	Minimum Zenith Distance	Maximum Altitude
Regulus.....	7:04 P.M.	27°.6	62°.4 (crosses south of zenith)
Spica........	10:21 P.M.	50°.7	39°.3 (crosses south of zenith)
Arcturus.....	11:12 P.M.	20°.4	69°.6 (crosses south of zenith)

FIRST-MAGNITUDE STARS VISIBLE IN LAT. 40° N. WHICH WILL CROSS MERIDIAN DURING FIRST HALF OF THE NIGHT OF MAY 6

WERE we to confine ourselves to stellar positions at meridional passage, obviously it would be a tedious task to master the topography of the heavens. What we desire is a scheme by means of which we can determine where we ought to look for a star or group of stars, given date, time, and latitude. Our first step is to develop a means of specifying position in the sky, which we have partly accomplished in defining altitude. Thus, altitude 90° means that the object is overhead; altitude 0°, that it is somewhere on the horizon; altitude 45°, that it is halfway between horizon and zenith. But unless the altitude is 90° or −90° we do not know its precise position; there still remains for us to specify the direction in which we ought to face. This could be accomplished by stating whether it is north, east, south, or west and by adding one of the points of the compass to each statement of altitude.

But how many of us are nautically inclined? Most of us are landlubbers and somewhat slow in comprehending such terms as "east-northeast" and "northeast-by-north." We shall find it much simpler to express position on the horizon in terms of the angular distance east or west of the north point; and this angle we call AZIMUTH, which is derived from the Arabic *as-sumūt*, meaning "the ways." We denote a direction due north by azimuth 0°; northeast, by azimuth 45° E.; east, by azimuth 90° E.; south, by azimuth 180°; and west, by azimuth 90° W. The correspondence between the more common points of the compass and azimuth is shown in the table.

Point of compass	N.	NE.	E.	SE.	S.	SW.	W.	NW.
Azimuth	0°	45° E.	90° E.	135° E.	180°	135° W.	90° W.	45° W.

Given the altitude and azimuth of a star, its position in the sky is determined; thus an altitude of 0° and azimuth $22\frac{1}{2}°$ W. implies that the object is on the horizon halfway between a point due north and one northwest. To find a star with an altitude of 30° and azimuth of 135° W., we face the southwest and locate the star in that region of the sky which is one-third of the way upward from the horizon to zenith.

IT IS not practical to tabulate the altitude and azimuth for every celestial object of interest for all latitudes, days, and hours. While the right ascension and declination of a star will remain essentially the same throughout our lifetime, altitude and azimuth are continually changing. In stellar catalogues, therefore, right ascension and declination only are specified; and it is left to us to convert these EQUATORIAL CO-ORDINATES, as they are termed, into altitude and azimuth, the HORIZON CO-ORDINATES.

By comparing the right ascension of a star with the sidereal time, we can determine either the time that has elapsed since the star crossed the meridian or the number of hours it will take to reach the meridian. For example, suppose it is January 4 at 10:00 P.M., which makes sidereal time 5:00. On consulting the table of first-magnitude stars, we find that Aldebaran (right ascension 4^h31^m) crossed the meridian about half an hour ago; on the other hand, Rigel (right ascension 5^h11^m) will cross in another 11 min. Since the stars travel from east to west, we call the interval of time that has elapsed since a particular star passed the meridian its HOUR ANGLE WEST, while the time it will take to attain the meridian is its HOUR ANGLE EAST. The hour angle, therefore, is the difference between sidereal time and the object's right ascension or, expressed in a formula,

Hour angle west (time since last meridional transit)	=	sidereal time MINUS right ascension;
Hour angle east (time until next meridional transit)	=	right ascension MINUS sidereal time.

In attempting the indicated subtraction, the subtrahend might exceed the minuend; so, to avoid a negative hour angle, we add 24 hr. to the subtrahend and then subtract—a procedure similar to that employed when we encountered the same difficulty in the conversion of sidereal time to hours past noon. In other words, we are permitted to add or subtract 24 hours in any sum or difference involving sidereal time. For example, at 10:00 sidereal time, Spica (right ascension 13^h21^m) has an hour angle of 10 *plus* 24 *minus* $13^h21^m = 20^h39^m$ W. and an hour angle of 13^h21^m *minus* $10 = 3^h21^m$ E. As indicated by the example, every star has an hour angle east and an hour angle west; and, to avoid confusion, we shall always take the smaller hour angle (at most, it can be 12 hr.), specifying whether it is east or west.

Given hour angle, declination, and latitude, the altitude and azimuth of the object are determinable. Unfortunately, the mathematical manipulations required for this final step are a trifle more complicated than the simple additions and subtractions we have so far encountered. Formulae for this problem are given in texts on spherical trigonometry;* but, for a fairly close determination of position, graphical methods may be utilized. Charts 1 and 2 may be regarded as replacing those formulae of spherical trigonometry; and, with reasonable care, the student may confine his error to not more than 2° of arc. We regard the substitution of these charts for

---

* These formulae may be written
$$\sin a = \sin d \sin l + \cos d \cos l \cos h,$$
$$\sin d = \sin a \sin l + \cos a \cos l \cos A,$$
where d is declination; h, hour angle; l, latitude; a, altitude; and A, azimuth.

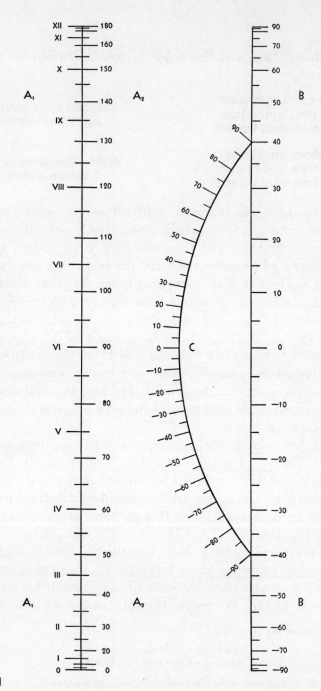

CHART 1

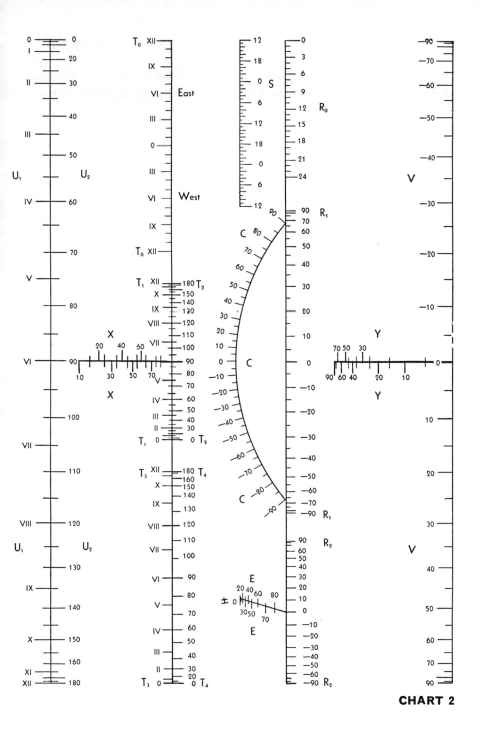

CHART 2

more or less involved mathematics as legitimate, for two reasons: *imprimis*, they require no mathematical knowledge in their manipulation, and students not versed in mathematics are thereby enabled to solve a variety of interesting problems in astronomy; *secundus*, approximate results are more quickly obtained than by the formulae of spherical trigonometry.

Before considering the charts, we take the liberty of suggesting that the reader spend a little time familiarizing himself with the steps involved in determining sidereal time and hour angle. With practice, the calculation of the hour angle will seem as simple as reckoning the interval of time from 10:00 A.M. to 3:00 P.M.; in fact, the two problems would be identical were we accustomed to the use of clocks reading 24 hr. instead of 12.

CHART 1 is for Lat. 40° N.; and, with the aid of a ruler or any straightedge (a taut thread will do very well), azimuth and altitude are quickly obtained for this latitude, given hour angle and declination. Despite the fact that the inhabitants of New York City, Philadelphia, Pittsburgh, Chicago, Kansas City, and San Francisco are sufficiently close to this parallel of latitude to utilize Chart 1 in locating celestial objects, we hasten to admit that 40° N. is not the only latitude and call attention to Chart 2, which is applicable to *any* latitude. Let us begin with Chart 1 as a basis for our study and imagine ourselves as being on the fortieth parallel north.

This chart consists of two graduated parallel scales A_1, or A_2, and B and an arc of a circle, C. Roman numerals 0, I, II, up to XII on A_1 scale represent hours, while the Arabic numbers on scales A_2, C, and B indicate degrees. Each point on these scales should be regarded as representing a certain number of degrees or hours; e.g., a point on the scale midway between marking 20 and 30 corresponds to 25°, and a point midway between VIII and IX corresponds to 8^h30^m. If a straight line is drawn on the chart so as to intersect scales A_1, B, C, then the points at which this straight line cuts A_1, B, and C define three quantities. The chart is so constructed that, if the intersection on the A_1 scale represents hour angle and that on C scale declination, then the point where the straight line cuts scale

THE SKY

B indicates the object's altitude for latitude 40° N. In other words, given hour angle and declination, to solve the problem of determining altitude for latitude 40° N., we join by a straight line the points on A_1 and C scales representing hour angle and declination; and, at this line's intersection with B scale, we read off the required altitude.

Now, knowing the altitude and declination, it is possible to find the object's azimuth by means of this selfsame chart. This time we regard the B scale as representing declination; the C scale, altitude (just the reverse of the previous problem); and by joining with a straight line the points on B and C corresponding to the assigned declination and altitude, at this line's intersection with the A_2 scale we read off the azimuth of the object. There remains only the problem of the determining whether this azimuth is east or west, and this is accomplished by the direction of the hour angle, the rule being

If hour angle is east then azimuth is east;
If hour angle is west then azimuth is west.

To avoid mutilating the page by actually drawing a straight line on the chart, the same results can be obtained by placing a ruler so that its edge cuts the correct markings and reading off the intersections on the scales directly. For ready reference, we summarize, in the accompanying table, the two correspondences obtained by this procedure.

Scales	A_1	A_2	B	C
Correspondence 1	Hour angle		Altitude	Declination
Correspondence 2		Azimuth	Declination	Altitude

The chart is not limited to the two types of problems we have mentioned; in fact, given any two of the quantities listed in correspondence 1, the third is determinable, and the same property holds for correspondence 2; furthermore, it is not necessary that the celestial object be a star. *The American Ephemeris and Nautical Almanac* contains not only the right ascensions and declinations of many stars but also those for the planets, moon, and sun. Though we shall postpone for the present the problem of locating the planets, the reader, if he wishes, may consult the *Nautical Almanac* for their equatorial co-ordinates and derive their altitude and azimuth with the aid of Charts 1 and 2. Though the declination of the sun varies from day to day, for the same dates in subsequent years it is essentially the same, and we may therefore conveniently tabulate its value (see Table 5).

Day	Jan.	Feb.	Mar.	Apr.	May	June	July	Aug.	Sept.	Oct.	Nov.	Dec.
1	−23°0	−17°2	− 7°8	4°4	14°9	22°6	23°1	18°1	8°5	− 3°0	−14°3	−21°7
6	−22.5	−15.8	− 5.8	6.3	16.6	23.1	22.6	16.8	6.6	− 4.9	−15.8	−22.4
11	−21.9	−14.2	− 3.9	8.1	17.7	23.3	22.2	15.4	4.8	− 6.8	−17.3	−23.0
16	−21.0	−12.5	− 1.9	10.0	19.0	23.4	21.5	13.9	2.8	− 8.7	−18.6	−23.3
21	−20.0	−10.7	0.0	11.8	20.1	23.4	20.6	12.3	0.9	−10.5	−19.8	−23.4
26	−18.8	− 8.9	2.0	13.4	21.0	23.4	19.6	10.6	−1.0	−12.3	−20.8	−23.4
31	−17.5		4.0		21.8		18.4	8.8		−14.0		−23.1

TABLE 5

DECLINATION OF SUN IN DEGREES

The following examples illustrate many applications of Chart 1. In all cases the latitude is 40° N.

a) *Type of problem:* Given date, time, right ascension, and declination; find altitude and azimuth of object.

Procedure:
1. Compute sidereal time (see Table 4).
2. Find hour angle, noting whether east or west.
3. Using Chart 1 with correspondence 1, pass straightedge through hour angle on A_1 scale and declination on C scale. At intersection on B scale read object's altitude.
4. Using Chart 1 with correspondence 2, pass straightedge through declination on B scale and altitude on C scale, reading azimuth on A_2 scale. Direction of azimuth, i.e., whether east or west, is same as that of hour angle.

Numerical example:
Date June 22, time 8:10 P.M., object Vega, right ascension 18^h34^m, declination 38°7; find altitude and azimuth and locate star.
1. Sidereal time, 8:10+6:10 = 14:20.
2. Hour angle, 18:34−14:20 = 4:14 E. (i.e., star will cross meridian 4^h14^m after specified time).
3. By passing straightedge through 4:14 on A_1 scale and 38°7 on C scale, intersection on B scale indicates altitude 42°.
4. By passing straightedge through 38°7 on B scale and 42° on C scale, intersection on A_2 scale indicates azimuth 70° E.

THE SKY

On June 22 at 8:10 p.m. in Lat. 40° N. the star Vega is a trifle north of east and about halfway between the horizon and zenith.

Numerical example:
Date December 5, time 9:25 p.m., object Deneb, right ascension 20^h39^m, declination 45°.0; find altitude and azimuth.
1. Sidereal time, $9:25+17:01=26:26=2:26$.
2. Hour angle, $2:26+24:00-20:39=5:47$ W. (Subtracting 2:26 from 20:39 would give hour angle 18:13 E.; but this is greater than 12 hr., and hence we compute the hour angle west, which is $24-18:13=5:47$.)
3. By correspondence 1, altitude is 29°.
4. By correspondence 2, azimuth is 54° W.

b) *Type of problem:* Given declination; find azimuth of point where object rises and of point where object sets.

Procedure: Since object is on horizon, altitude is 0°. Using Chart 1 with correspondence 2, pass straightedge through assigned declination on B scale and altitude 0° on C scale, reading object's azimuth at intersection on A_2 scale; azimuth is east if object is rising, west if setting.

Numerical example:
Find rising point of sun on June 22.
According to Table 5, declination of sun on June 22 is 23°.4.
By passing straightedge through 23°.4 on B scale and 0° on C scale, at intersection on A_2 scale we find azimuth to be 59° E. (Had the problem been to find the setting point of the sun, the procedure would have been the same and azimuth would have been 59° W.)

c) *Type of problem:* Given declination; find length of time above horizon.

Procedure: The length of time the sun, planets, or stars are above the horizon in mean solar time is approximately twice the hour angle at rising. Chart 1 is used with correspondence 1, the straightedge being passed through altitude 0° on B scale and assigned declination on C scale. The intersection with A_1 scale gives hour angle at rising. (For a star this quantity is equal to the interval of time it will take to reach meridian in sidereal time units and, hence, is the approximate time in mean solar units.)

Numerical example:
Find duration of sunlight on December 22.
From Table 5 declination of sun is $-23°.4$.
Pass straightedge through 0° on B scale and through $-23°.4$ on C scale. At intersection on A_1 scale we read hour angle at rising is 4^h35^m; therefore duration of sunlight on December 22, Lat. 40° N. is $4^h35^m \times 2 = 9^h10^m$.

d) *Type of problem:* Given right ascension, declination, date; find time of rising and time of setting.

Procedure: As in problems of type (c), determine hour angle at rising or setting by passing straightedge through 0° altitude on B scale and declination on C scale to read hour angle on A_1 scale. If object is rising, subtract hour angle from right ascension, adding 24 hr. if necessary; if object is setting, add hour angle to right ascension. The result is sidereal time, and local time is then obtained by subtracting the proper entry given in Table 4.

Numerical example:
Determine time of rising and time of setting for Pollux on August 1.
    According to Table 3, right ascension is 7^h40^m, declination $28°.2$.
    By joining straightedge at $0°$ on B scale with $28°.2$ on C scale, at intersection with A_1 the hour angle is 7^h50^m E. if star is rising, 7^h50^m W. if setting.
    Sidereal time at rising $= 7:40+24-7:50 = 23:50$.
    Sidereal time setting $= 7:40+7:50 = 15:30$.
    Hours past noon at rising $= 23:50-8:44 = 15:06 = 3:06$ A.M.
    Hours past noon at setting $= 15:30-8:44 = 6:46 = 6:46$ P.M.

e) *Type of problem:* Given date, time, altitude, and azimuth; find right ascension and declination.
Procedure:
1. Compute sidereal time.
2. Using correspondence 2, pass straightedge through azimuth on A_2 scale and altitude on C scale; at intersection on B scale read object's declination.
3. Using correspondence 1, pass straightedge through altitude on B scale and declination on C scale, reading hour angle on A_1 scale.
4. If azimuth is east, hour angle is east and is added to sidereal time to obtain right ascension; if azimuth is west, hour angle is west and is subtracted from sidereal time (adding 24 hr. if necessary) to find right ascension.

Numerical example:
In Philadelphia on January 23 at 9:45 P.M., E.S.T., a first-magnitude star was estimated to have an altitude of $30°$ and azimuth of $100°$ E. Identify the star.
1. Sidereal time, $9:45+20:15 = 30:00 = 6:00$.
2. By passing straightedge through $100°$ an A_2 scale and $30°$ on C scale, declination at intersection on B scale is $12°$.
3. By passing straightedge through $30°$ on B scale and $12°$ on C scale, hour angle at intersection on A_1 scale is 4^h10^m E.
4. Right ascension of star is $6:00+4:10 = 10^h10^m$.

By Table 3 we discover that the star probably was Regulus (right ascension 10^h4^m, declination $12°.4$).

f) *Type of problem:* Given declination; determine hourly position of object in sky.
    Procedure: Using correspondence 1, find altitude for different hour angles by passing straightedge through declination on C scale and through hour angles 0, I, II, and so on up to and including XII, noting at each step the corresponding altitude on B scale.

THE SKY

Using correspondence 2, determine azimuth for each altitude by passing straightedge through declination on B scale and computed altitudes on C scale, reading for each one the azimuth on A_2 scale.

Numerical example:

Find the circle described by the sun on June 22. The declination of the sun for this date is 23°.4.

Following the procedure outlined above,

Hour angle east or west..........	0	I	II	III	IV	V	VI	VII	VIII	IX	X	XI	XII
Altitude.........	73°	70°	60°	49°	37°	26°	15°	4°	−6°	−14°	−20°	−25°	−27°
Azimuth east or west.........	180°	145°	115°	100°	90°	80°	72°	63°	54°	42°	29°	15°	0°

At hour angle 0, object is on meridian; at hour angle XII, it is on lower meridian.

Analogously, it can be demonstrated with this chart that stars with declinations between 50° and 90° are always above the horizon, while stars with declinations between −50° and −90° are never visible in latitude 40° N.

WE NOW turn to Chart 2, which permits us to solve the previous problems for practically any latitude on the surface of the earth. The chief features of this chart are four vertical graduated scales labeled, respectively, from left to right (U_1 or U_2), (T_0, T_1, T_2, T_3, or T_4), (R_0, R_1, or R_2), and V, together with two short horizontal scales, X and Y, and an arc of a circle, C. For the present we shall ignore, not only the scales S and E, but also the calibrations on the series of T and R scales, regarding the graduations merely as a means of locating a point on the T and R lines. It is possible to calibrate the lines T and R for any assigned latitude, excluding the equator, so that the resulting scales, taken in conjunction with the graduated circular arc C, constitute a chart similar in all respects to Chart 1, except that it would apply to this new latitude. As a matter of fact, as it now stands, scales T_1 or T_2, R_1, and C form a chart analogous to Chart 1 for latitude $66\frac{1}{2}°$ N., the arctic circle, scales T_1 and T_2 corresponding to A_1 and A_2, and R_1 to B. Therefore, by reading T_1 for A_1, T_2 for A_2, and R_1 for B, we can follow the procedure outlined in the previous problems to solve them for the arctic circle.

To calibrate line T for any *other* latitude between 10° N. and 90° N., i.e., to find the location of 0, I, II, etc., as well as 0, 10, 20, etc., we utilize scales U_1, or U_2, and X. First pass a straightedge through the assigned latitude on scale X and through 0 on U_1 scale. The intersection on T line gives the location of 0 for the particular latitude. The operation is repeated; but this time the straightedge is passed through the assigned latitude on X scale and through I on U_1 scale, which locates I on the T line. The procedure is repeated through II, III, IV, and so on until all the markings on U_1 and U_2 are transferred in this fashion to the T line. In like manner, the 0, 10, 20, etc., for the R line are obtained by passing a straightedge through the assigned latitude on Y scale and through the numbers 0, 10, 20, and so on, on V scale in turn, marking at each stage the intersection on the R scale. If the reader desired, he could copy lines T and R and arc C on tracing paper and calibrate them in this manner for his own latitude.

In order to solve problems, it is not essential to calibrate lines T and R, for, by means of scales X and U_1 or U_2, not only can we locate position on line T but also, given a point on line T, we can find its significance. The same property applies to the R line in conjunction with the Y and V scales. For example, suppose we have been given the hour angle, declination, and latitude and wish to find the object's altitude. We pass our straightedge through the hour angle on U_1 scale and through the latitude on X scale, noting the place where the straightedge cuts the T line, the T *point* as we shall term it. The straightedge is next passed through the T point and through the declination on C scale to intersect at an R *point* on the line R. Finally, by placing the straightedge through the R point and through the latitude on Y scale, at its intersection with V scale we read the object's altitude.

In the interests of brevity, we shall abbreviate the description of these operations. Let us therefore denote the fact that we are reading the U_1 scale as hour angle by writing U_1 (*hour angle*); similarly, X (*latitude*) will mean that the X scale gives latitude. Furthermore, we shall indicate the operation of placing a straightedge

THE SKY

through hour angle on U_1 scale and through latitude on X scale to determine a T point by writing:

$$U_1 \text{ (hour angle)} + X \text{ (latitude)} \rightarrow T \text{ point}.$$

The remaining steps in determining altitude from hour angle, latitude, and declination then become in this notation:

$$T \text{ point} + C \text{ (declination)} \rightarrow R \text{ point};$$
$$R \text{ point} + Y \text{ (latitude)} \rightarrow V \text{ (altitude)}.$$

For ready reference we outline the steps taken in using Chart 2 for all latitudes, except those within 10° of the equator.

Case A.—Given latitude, hour angle, declination; find altitude.

$$U_1 \text{ (hour angle)} + X \text{ (latitude)} \rightarrow T \text{ point};$$
$$T \text{ point} \quad + C \text{ (declination)} \rightarrow R \text{ point};$$
$$R \text{ point} \quad + Y \text{ (latitude)} \quad \rightarrow V \text{ (altitude)}.$$

Case B.—Given latitude, declination, altitude; find azimuth.

$$V \text{ (declination)} + Y \text{ (latitude)} \rightarrow R \text{ point};$$
$$R \text{ point} \quad + C \text{ (altitude)} \rightarrow T \text{ point};$$
$$T \text{ point} \quad + X \text{ (latitude)} \rightarrow U_2 \text{ (azimuth)}.$$

Case C.—Given latitude, altitude, and azimuth; find declination.

$$U_2 \text{ (azimuth)} + X \text{ (latitude)} \rightarrow T \text{ point};$$
$$T \text{ point} \quad + C \text{ (altitude)} \rightarrow R \text{ point};$$
$$R \text{ point} \quad + Y \text{ (latitude)} \rightarrow V \text{ (declination)}.$$

Case D.—Given latitude, altitude, and declination; find hour angle.

$$V \text{ (altitude)} + Y \text{ (latitude)} \quad \rightarrow R \text{ point};$$
$$R \text{ point} \quad + C \text{ (declination)} \rightarrow T \text{ point};$$
$$T \text{ point} \quad + X \text{ (latitude)} \quad \rightarrow U_1 \text{ (hour angle)}.$$

Case E.—Alternate method of finding hour angle, given latitude, declination, and altitude.

$$V \text{ (altitude)} + Y \text{ (declination)} \rightarrow R \text{ point};$$
$$R \text{ point} \quad + C \text{ (latitude)} \quad \rightarrow T \text{ point};$$
$$T \text{ point} \quad + X \text{ (declination)} \rightarrow U_1 \text{ (hour angle)}.$$

It will be noted in cases A, B, C, and D that the X and Y scales correspond to latitude and that these two scales are calibrated from $+10$ to $+90$, i.e., Lat. 10°–90° N. But we are not confined to the northern hemisphere in the application of Chart 2. If we denote latitudes south of the equator by negative numbers, our chart is still valid, provided we change the signs of the numbers on the X, Y, and C scales. In other words, we are permitted to replace the numbers 10, 20, 30, etc., on the X and Y scales by -10, -20, -30, etc., so long as we also change the sign of all the numbers on scale C, positive numbers becoming negative and negative numbers positive.

Chart 2 includes all latitudes, except a small region near the earth's equator. Rather than omit the equator entirely, we have included a special equatorial section in scales T_3, T_4, E, and R_2. The scales T_3 and T_4 correspond to the A_1 and A_2 scales of Chart 1, the R_2 scale to the B, and the E to C, the numbers on the E scale being taken with either a positive or a negative sign. This equatorial chart is used in precisely the same manner as Chart 1. For the equator, therefore, in terms of the adopted notation:

T_3 (hour angle) $+E$ (declination)$\rightarrow R_2$ (altitude) ;
R_2 (declination)$+E$ (altitude) $\rightarrow T_4$ (azimuth) .

Though we have so far insisted that the reader be able to compute hour angle, given sidereal time and right ascension, we include in the remaining scales T_0, S, and R_0 of Chart 2 a means of solving this problem graphically. We symbolize the procedure by

S (sidereal time)$+R_0$ (right ascension)$\rightarrow T_0$(hour angle) ,

and urge that the chart be used only as a check.

First of all we shall work out one or two numerical examples and then, as a challenge to the student, suggest a few problems.

a) *Type of problem:* Given date, time, latitude, right ascension, and declination; locate object, i.e., find altitude and azimuth.

Procedure:
 1. Compute sidereal time and hour angle.
 2. Following case A, find altitude.
 3. Following case B, find azimuth.

Numerical example:
Find altitude and azimuth of Vega (right ascension $18^h 34^m$, declination $38°.7$) in London, England (Lat. $51°.5$ N.), for June 22 at 8:10 P.M., G.M.T.
 1. Sidereal time, $8:10+6:10=14:20$.
    Hour angle, $18:34-14:20=4^h 14^m$ E.

THE SKY

2. Straightedge through hour angle 4:14 on U_1 scale and latitude 51°.5 on X scale intersects T_2 scale at 33°. Straightedge through 33° on T_2 scale and declination 38°.7 on C scale intersects R_1 scale at 57°. Straightedge through 57° on R_1 scale and latitude 51°.5 on Y scale intersects V scale at altitude 45°.

3. Straightedge through 38°.7 on V scale and latitude 51°.5 on Y scale intersects R_1 scale at 50°. Straightedge through 50° on R_1 scale and altitude 45° on C scale intersects T_2 scale at 70°. Straightedge through 70° on T_2 scale and latitude 51°.5 on X scale intersects U_2 scale at azimuth 80°.

Answer: Altitude, 45°. Azimuth, 80° E.

Numerical example:

Find altitude and azimuth of Acrux (right ascension 12^h22^m, declination $-62°.6$), for Buenos Aires (Lat. 35° S.), at midnight on December 24.

1. Sidereal time = 12:00+18:16 = 6:16.
   Hour angle = 12:22−6:16 = 6:06 E.
   (Since Buenos Aires is in the southern hemisphere, we change signs of numbers on X, Y, and C scales.)
2. Following case A, altitude is 30°.
3. Following case B, azimuth is 135° E.

b) *Type of problem:* Given declination and latitude; find azimuth of rising point and of setting point.

Procedure: Since the altitude is 0°, we may follow case B to determine azimuth. Scales T_4, E, and R_2 may also be used for this purpose, for

R_2 (declination)$+E$ (latitude north or south)$\to T_4$ (azimuth),

provided the object is on the horizon.

Numerical example:

Find rising point of sun in Miami, Florida (Lat. 25°.7 N.), on December 22. Declination of sun on this date is −23°.4.

By passing straightedge through −23°.4 on R_2 scale and 25°.7 on E scale, at intersection on T_4 scale azimuth is 116° E.

c) *Type of problem:* Given latitude and declination; find length of time above horizon.

Procedure: Find hour angle at rising by following case D. Incidentally, the steps can be simplified as follows:

R_1 (0 point)$+C$ (declination)$\to T$ point ;
T point $\quad +X$ (latitude) $\quad \to U_1$ (hour angle at rising) .

Twice the hour angle is equal to the approximate time above the horizon.

Numerical example:

Find duration of sunlight in Los Angeles, California (Lat. 34° N.), on June 22. Declination of sun for that date is 23°.4.

Hour angle at rising is 7^h8^m; therefore duration of sunlight is 14^h16^m. (All statements in regard to southern California sunlight are taken at their face value.)

d) *Type of problem:* Given latitude, right ascension, declination, date; find time of rising and time of setting.

Procedure: As in problems of type (c), find hour angle at rising or setting. If object is rising, subtract hour angle from right ascension, adding 24 hr. if necessary; if setting, add hour angle to right ascension. The result is sidereal time, and local time is then obtained by subtracting the proper entry given in Table 4.

Numerical example:
Find time of rising of Acrux (alpha of the Southern Cross, right ascension 12^h22^m, declination $-62°\!.6$) in Key West, Florida (Lat. $24°\!.5$ N.), on April 1.

In this problem the T line must be extended a trifle in order to obtain the intersection with straightedge through 0 on R_1 scale and $-62°\!.6$ on C scale.

By joining this intersection with $24°\!.5$ on X scale, hour angle at rising is found to be 2^h E.

Sidereal time at rising $= 12:22 - 2:00 = 10:22$.
Local time at rising $= 10:22 - 0:43 = 9:39$ P.M.

e) *Type of problem:* Given date, time, altitude, and azimuth; find right ascension and declination.
Procedure:
1. Compute sidereal time.
2. Following case C, find declination.
3. Following case D, find hour angle.
4. If azimuth is east, then hour angle is east and is added to sidereal time to obtain right ascension; if azimuth is west, then hour angle is west and is subtracted from sidereal time (adding 24 hr. if necessary) to find right ascension.

Numerical example:
An inhabitant of New Orleans, Louisiana (Lat. 30° N.), writes that on September 1 at 8:15 P.M. he saw a bright star directly southeast at an estimated angular height (altitude) of about 30°. He would like to know its name.
1. His sidereal time is $8:15 + 10:46 = 19:01$.
2. Since azimuth is 135° E. and estimated altitude 30°, by following case C declination is $-16°$.
3. Following case D, hour angle is 2^h40^m E.
4. Right ascension, $2:40 + 19:01 = 21:41$.

Turning to Table 3, we find that there is no first-magnitude star near declination $-16°$ and right ascension 21:41. Former experiences of this nature direct us to the *Nautical Almanac*, from which we learn that on September 1, 1934, the planet Saturn had a right ascension of 21^h47^m and a declination of $-15°$. We politely inform our correspondent that he was looking at a planet and not at a star.

f) *Type of problem:* Given latitude and declination; determine hourly position.
Procedure:
1. Following case A, determine altitude for hour angles 0, I, II, and so on up to XII.
2. Following case B, compute the azimuth for each of these altitudes.

THE SKY

Numerical example:

Find hourly position of the sun at Lat. 40° N. and at the equator on June 22.

The hourly position of the sun for 40° N. has already been determined, using Chart 1. As a drill, the student may re-work this same problem, using the more general Chart 2.

At the equator, the previous procedure fails and, in its place, we employ the special scales T_3, T_4, E, and R_2. The altitude at different hour angles is found from:

$$T_3 \text{ (hour angle)} + E \text{ (declination)} \rightarrow R_2 \text{ (altitude)}.$$

The azimuth from these altitudes is then found from:

$$R_2 \text{ (declination)} + E \text{ (altitude)} \rightarrow T_4 \text{ (azimuth)}.$$

On June 22 the declination of the sun is 23°.4, and consequently its hourly positions at the equator are:

Hour angle east or west..........	0	I	II	III	IV	V	VI	VII	VIII	IX	X	XI	XII
Altitude.........	67°	62°	53°	40°	27°	13°	0°	−13°	−27°	−40°	−53°	−62°	−67°
Azimuth east or west..........	0°	30°	50°	60°	64°	66°	67°	66°	64°	60°	50°	30°	0°

By locating these points in the sky, the sun's path at the equator may be traced. If he cares to do this, the student will find that the sun comes up at right angles to the horizon and sets at right angles. In fact, *every* celestial object at the earth's equator crosses the horizon at right angles.

Supplementary problem:

On New Year's Day, after riding a terrific gale, a yachtsman on a cruise around the world finds himself becalmed. Not only is he ignorant of his position, but his magnetic compass has been thrown out of adjustment. While repairing the damage done by the storm, he observes the sun climb up in the sky. His chronometer indicates Greenwich Mean Time, and at 19:22 G.M.T., he finds that the sun is 8° above his horizon. In the next 2 hr. the sun's angular height increases steadily to an altitude of 30°. All he knows is that he is somewhere between the equator and the antarctic circle, and we proceed to locate him.

On January 1 the sun's declination is −23°.

According to the yachtsman's observations, the sun's altitude increased from 8° to 30° in a period of 2 hr. If we knew his latitude, we could follow case E and find the hour angle of the sun for altitudes 8° and 30°, and the difference between these hour angles would be 2 hr. But we do not know his latitude;

let us therefore try different latitudes, computing for each one the hour angles for altitudes 8° and 30° in the hope of discovering the particular latitude for which the hour angle difference is 2 hr. We follow case E (changing signs of X, Y, and C scales) and tabulate our results (see accompanying table). We examine the last column and discover that, at Lat. 40° S., the hour angle difference is close to 2 hr., and we infer that the yachtsman is near this parallel. Furthermore, at 40° S., the difference is only 2 min. *less* than 2 hr., while at 50° S. it is 24 min. greater, clearly indicating that he is a trifle north of 40° S. Let us try Lat. 41° S. (see table). There remains the problem of finding the yachtsman's longitude. On the basis of our successful trial with Lat. 41° S., at altitude 8° the sun was $6^h 38^m$ away from his meridian; in other words, his true solar time was $12:00 - 6:38 = 5:22$ A.M. On January 1 the equation of time (see Table 1) is -4 min., so that his mean

Assumed Latitude	Hour Angle for Altitude 8°	Hour Angle for Altitude 30°	Difference
70° S.	9:00 E	4:20 E	4:40
60°	7:40	4:36	3:04
50°	7:04	4:40	2:24
40°	6:36	4:38	1:58
30°	6:16	4:30	1:46
20°	6:00	4:28	1:32
10°	5:40	4:08	1:32
0°	5:24	3:50	1:34

Assumed Latitude	Hour Angle for Altitude 8°	Hour Angle for Altitude 30°	Difference
41° S.	6:38 E.	4:38 E.	2:00

solar or his local time was 5:26 A.M.; but, according to his chronometer, the time at Greenwich was 19:22, which would make his local time $5:26 + 24:00 - 19:22 = 10:04$ *ahead* of Greenwich Mean Time. Since an hour's difference in time corresponds to 15° difference in longitude, he is at Long. 151° E. of Greenwich. On consulting an atlas, we locate our yachtsman off the coast of Tasmania.

For him there is still the problem of steering his ship without the aid of a compass; but, knowing his latitude, he can find the azimuth or direction of the sun for, as outlined in case B, it is possible to compute azimuth, given latitude, altitude, and declination.

IF, IN reworking this problem, the student places the yacht in the interior of Australia, he should not be alarmed. We wish it clearly understood that, if Chart 2 is applied to problems in navigation, we will assume no responsibility, for, though it is entirely satisfactory for purposes of identifying celestial objects, its small scale does not permit of precise results.

Nevertheless, our supplementary example illustrates a fundamental problem in navigation, namely, the determination of posi-

THE SKY

tion from the observation of two altitudes of the sun or any other celestial object. "Shooting the sun" in nautical phraseology signifies the measurement of its altitude by means of a sextant. Though we cannot here go into the numerical calculations, we *can* outline the basic theory.

From tables given in the *Nautical Almanac* it is possible to determine with a high degree of accuracy the particular point on earth where the sun's center is at the zenith, provided Greenwich Mean Time is known. If the navigator finds the sun's altitude to be 90°, i.e., at the zenith, his position would be established; but it is improbable that he will be so situated and, in general, the sun's altitude will be less than 90°, the amount by which it differs being equal to the angular distance on the earth from the point directly below the sun. For example, an altitude of 70° for the sun implies that the observer is 20° (about $20 \times 69 = 1,380$ statute miles or $20 \times 60 = 1,200$ nautical or geographical miles) away from the point where the sun is at the zenith. This angle may be measured in any direction, so that a single observation of the altitude of the sun, or, for that matter, of any celestial object, merely locates the ship somewhere on a circle, the center of the circle being the known point on the earth where, at the instant of the observation, the sun's center is directly overhead. A second observation of the sun's altitude at a later time yields a second circle of possible positions. If the ship has not moved in the interval, one of the two points where the second circle cuts the first is its position, the two possible intersections of the circles usually being so far apart that there is no question as to the proper selection. It is unlikely, of course, that the ship will remain stationary; and the navigator must mathematically move his circle of possible positions along with the vessel, the distance and direction traveled in the interval between the two observations being obtained from the log (a device for showing speed) and the compass.

When cloudy weather prohibits astronomical observations, the ship's position is found by the less reliable method of "dead reckoning," that is, by log and compass, the navigator's ever present hope being that he will have an opportunity to check or modify his calculations by an astronomical observation.

In connection with this problem, there is one correction that both navigator and astronomer must apply to observations of altitude or position of celestial objects. If a coin is placed in a tumbler of water, the position it appears to occupy will not be its actual position. This phenomenon, termed REFRACTION, also occurs in our atmosphere. Light entering the earth's atmosphere from space is bent downward, so that any celestial object appears to be at a greater altitude than it would be were it observed from an earth without an atmosphere. This refraction is greatest for objects near the horizon, decreasing rapidly from about $\frac{1}{2}°$ at altitude 0° to less than $\frac{1}{10}°$ at altitude 10°, 1 min. at altitude 45° to 0° at zenith. When we see the sun, of which the angular diameter is about $\frac{1}{2}°$, just clear the horizon, we know that, lacking earth's atmosphere, it would be just about to make its appearance. In this manner, refraction prolongs the period the sun is above the horizon; and in latitude 40° the duration of sunlight is thus increased approximately $5\frac{1}{2}$ min. Furthermore, when near the horizon, both sun and moon appear distorted, because the "edge" nearest the horizon is refracted more than the uppermost portion.

WE NOW suggest a few problems for the reader. First, he should find the rising point of the sun in his latitude for some convenient morning and verify his results by direct observation. If this problem is too difficult, he may modify it by substituting the word "setting" for "rising." In this connection, he might work out the hour angle at setting and then determine the sun's altitude at 1 hr. past sunset. By solving this selfsame problem for the equator, the student will be able to answer for himself the question, "Why is the duration of twilight shortest at the equator?"

As a final test, he should wait for a clear night and find the altitude and azimuth of each star in Table 3 for his particular latitude and any desired time, verifying his results for stars with positive altitude by seeing whether there *is* a bright star in each of the determined directions. With the many stars visible in the sky, the

THE SKY

student may occasionally wonder, after painstakingly following the schemes here outlined, whether the star he thinks he located is, after all, the one he sought. For positive identification, he ought to consult a star map, find the star in question on the map and see if the arrangement of the nearby stars on the map corresponds to that of the region of the sky he is examining.

IT IS convenient at this point to introduce the accepted mode of classifying stars as to their apparent brightness. This is done by assigning to each star a number called its MAGNITUDE. The scheme dates back to the time of Hipparchus, who selected about twenty of the brightest stars and called them first-magnitude stars. The faintest stars visible to the naked eye he classified as of sixth magnitude, and all the other stars visible to the eye fell in magnitude classes 1–6, the fainter the star the greater its magnitude.

Because the twenty brightest stars vary considerably in intensity, the *average* brightness of these stars is taken as the definition of magnitude 1, in order to satisfy the demands of precise magnitude determinations. By measurement it is found that the faintest stars visible to the eye are about a hundredth as bright as this average, or magnitude 1; and hence, retaining in essence the scheme of Hipparchus, sixth magnitude is defined as one-hundredth the brightness of first magnitude. The magnitudes 2, 3, 4, and 5 are then fixed so that

> Magnitude 1 is 2.512 times as bright as magnitude 2.
> Magnitude 2 is 2.512 times as bright as magnitude 3.
> Magnitude 3 is 2.512 times as bright as magnitude 4.
> Magnitude 4 is 2.512 times as bright as magnitude 5.
> Magnitude 5 is 2.512 times as bright as magnitude 6.

Consequently, magnitude 1 is $(2.512)^2$* times as bright as magnitude 3, or $(2.512)^5$* times as bright as magnitude 6. If the student takes the trouble to multiply 2.512 by itself five times, the result will be approximately 100; in fact, this property determined the

* $(2.512)^2$ is mathematical shorthand for 2.512×2.512.
  $(2.512)^5$ is mathematical shorthand for $2.512 \times 2.512 \times 2.512 \times 2.512 \times 2.512$.

number to be selected, for it was previously stated that a sixth-magnitude star is one-hundredth as bright as a first.

The scheme is extended in both directions. Stars not visible to the unaided eye have magnitudes greater than 6; on the other hand, celestial objects brighter than magnitude 1 have magnitudes less than 1, i.e., magnitude 0 is 2.512 times as bright as magnitude 1, magnitude -1 is 2.512 times as bright as magnitude 0 and $(2.512)^2$ times as bright as magnitude 1, and so on through the negative numbers. Even smaller steps than unity are employed, e.g., magnitude 4.5 for a star implies that its brightness is halfway along the scale between magnitudes 4 and 5, but these refinements need not concern us.

A word in passing—it will be noticed that in the table of the twenty brightest stars there is no star with a magnitude exactly 1; yet they are always known as the first-magnitude stars.

DESPITE the many references to the "countless" stars visible to the eye, we list the number brighter than a given magnitude. With exceptional eyesight the limit of visibility is between magnitudes 6 and 7, making, at most, about 10,000 stars within the reach of the eye; but of these, only about half are above the horizon at any one time, and a large percentage will be at such low altitudes that the atmosphere will considerably diminish their brightness. We may therefore say that the number of stars actually visible in the sky at any time by those with very good eyesight is in the region of 3,000.

Magnitude	Total Number of Stars Brighter than Listed Magnitude
2	41
3	138
4	530
5	1,620
6	4,850
7	14,300
10	324,000
15	32,000,000 (est.)
20	1,000,000,000 (est.)

In order to compare stars, we tabulate the ratio in brightness and magnitude difference:

To illustrate, we note in Table 3 that the magnitude of the brightest star, Sirius, is -1.6, while the faintest first-

Magnitude difference	1	2	3	4	5	6	10
Ratio in brightness	2.5	6.3	16	40	100	251	10,000

magnitude star, Regulus, is 1.3. The difference in magnitude is 2.9, or about 3; so we conclude that Sirius is almost sixteen times as bright as Regulus.

On this same scale the magnitude of the sun would be -26.7. If we wish to make a comparison of the brightness of the sun and that of Sirius, we note that the magnitude difference is about 25; and, since every 5 mag. correspond to a factor of 100 in brightness, the sun is about 10,000,000,000 (10^{10}) times as bright as our brightest star. From this number, one might compute the length of time required to obtain a good coat of tan, using merely the light of Sirius.

The full moon is about as bright as a candle placed 30 ft. away and has a magnitude of approximately -11.2; the combined light of nearly 1,600,000 full moons would be required to make day and night equally bright. This figure is indicative of how the human eye accommodates itself to changes in illumination, for it is possible to read a newspaper with only the full moon as a source of light.

Although there are catalogues available which include the right ascension, declination, and magnitude of all the stars visible to the eye, we do not insist that the reader compute their altitude and azimuth for his particular latitude and locate every one of them. We shall merely assist him in his identification of the simpler groups of stars.

A GLANCE at the stars makes it obvious that they are not distributed uniformly in the sky. Certain groupings of the comparatively brighter ones are noticed; and, joining them with imaginary lines, we form dippers, crosses, *W*'s, squares, and so forth in the heavens. Nor is this an idle pursuit, for it helps us to remember the stars and identify them. The stars were divided into groups, or CONSTELLATIONS, by the ancients, who attributed to them certain names now famous in mythology. In a few cases, by using the imagination, there still may be found some resemblance to the object from which a certain group derives its name; but, in general, no likeness can be seen nor can any reason be assigned for the name. Lacking instruments to determine precise position, the early Chinese, Babylonians, and Egyptians used these configurations for remembering the stars.

In a document attributed to Eudoxus (409–326 B.C.) there is a description of many of the constellations as they are known today. Ptolemy, in his star catalogue, divided the stars into forty-eight constellations—twelve along the ecliptic, twenty-one to the north, and fifteen to the south. But neither Eudoxus nor Ptolemy was the originator of this division of the skies. Inscriptions on monuments and tablets, as well as references in earlier writings, indicate that their work was based on observations thousands of years old. The absence of the elephant and the crocodile in the list indicates that they did not originate in India or Egypt, and all the evidence is said to point to the Euphrates Valley. As far as the stars visible in this locality are concerned, the division of the sky into constellations is believed to have been practically completed before 3000 B.C.

For almost fifteen centuries it was considered lèse-majesté to add to Ptolemy's list, but in 1601 Tycho Brahe added two constellations. His successors followed his lead, particularly in the region not visible to Ptolemy (the neighborhood of the south celestial pole), so that the number now generally recognized totals eighty-eight.

IN OLDER star maps we find considerable disagreement as to the boundaries and, in some cases, the designations of the constellations. It is amusing to read in former texts that "Aries should not have a horn in Pisces and a leg in Cetus; 51 Camelopardali might with propriety be extracted from the eye of Auriga and the ribs of Aquarius released from 46 Capricorni." For this reason the International Astronomical Union in 1922 fixed the boundaries of the constellations, so that it is now possible to specify them precisely in terms of right ascension and declination.

The designation of stars has likewise undergone great changes. The oldest method was to describe the object's position relative to the imaginary figure from which the constellation derived its name; thus a star was "in the Sword of Orion" or "at the tip of the Horn of the Bull," which accounts in some measure for the fantastic figures given in many star maps. In the translation of the names of

THE SKY

the stars we often discover this mode of designating location, as is evident from Table 3. The names of the brighter stars are of Greek, Latin, or Arabic origin; but, although the Arabs named many of the fainter stars, only some fifty star names in all are now in common use.

In the year 1603, Bayer introduced in his *Uranometria* one of our present means of designating stars—the name of the constellation prefixed by a Greek letter. Thus, Sirius was rechristened, or, strictly speaking, christened, α Canis Majoris and, for the benefit of those unfamiliar with the Greek alphabet, we list here the letters in order. In general, the stars are lettered in about their order of magnitude, α being the brightest, β the next brightest, and so on; but there are notable exceptions, e.g., the stars of the Dipper are lettered in the order in which they occur. When the Greek letters are exhausted, the Roman letters are employed; but even this is not sufficient, and we find stars designated merely by Arabic numerals or catalogue numbers.

α alpha	ι iota	ρ rho
β beta	κ kappa	σ sigma
γ gamma	λ lambda	τ tau
δ delta	μ mu	υ upsilon
ε epsilon	ν nu	φ phi
ζ zeta	ξ xi	χ chi
η eta	ο omicron	ψ psi
θ theta	π pi	ω omega

WE SHALL describe rather briefly some of the more easily recognizable constellations. Small maps of different regions of the sky are included here, the right ascension and declination center of the map being listed so that the student may determine the position the group occupies in his sky at any time. An arrow drawn on the map to indicate the direction of the north celestial pole enables him to find the orientation of the group; i.e., to compare a particular map with the sky, first determine the altitude and azimuth of the center of the map (hold the map about 5 inches in front of one eye with center in this direction, so that the arrow points to the north celestial pole), then the stars in the sky and their counterparts on the maps will be approximately in the same line with the eye.

In this connection we mention the "Stellarscope," a device which will help the student in the use of this book. Maps similar to those given here are photographed on motion-picture film and viewed

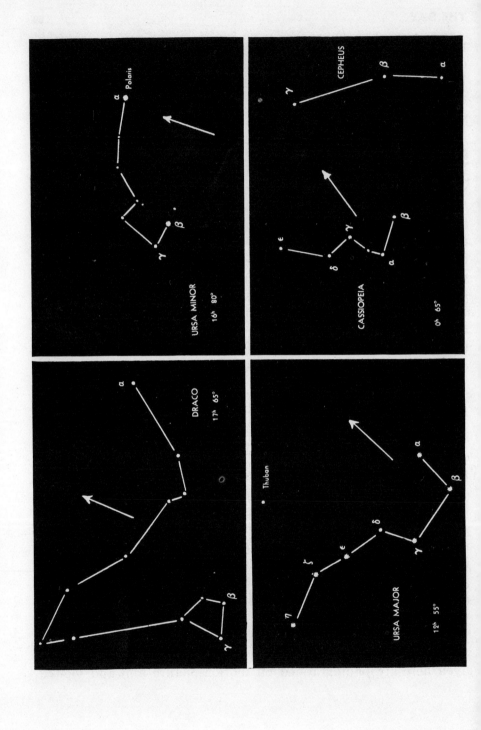

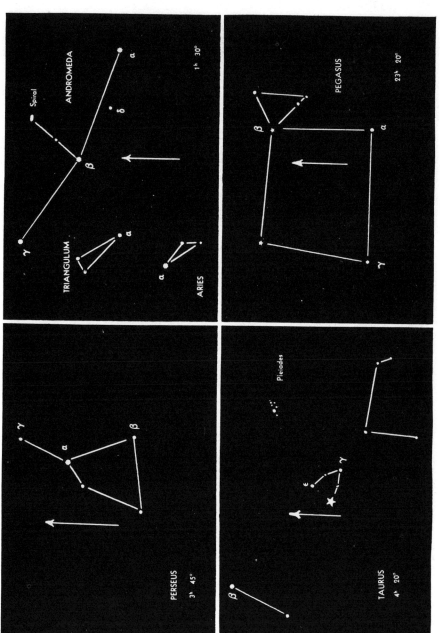

STAR MAP 2

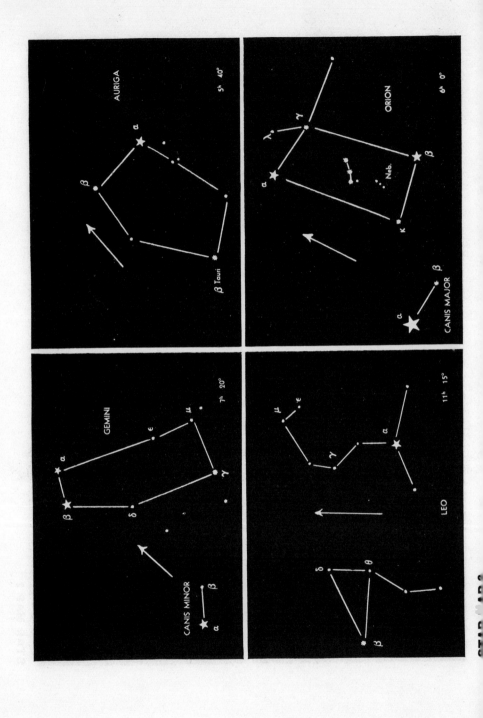

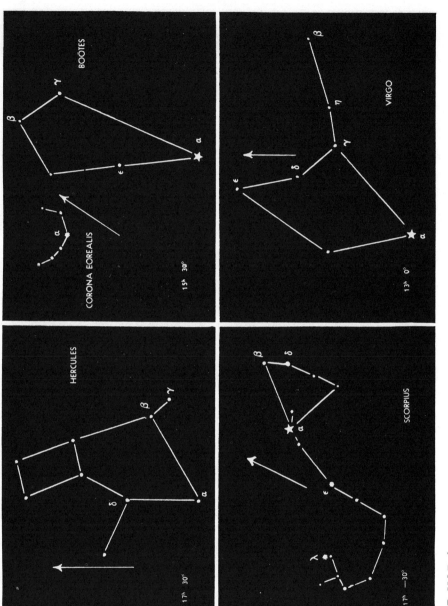

STAR MAP 4

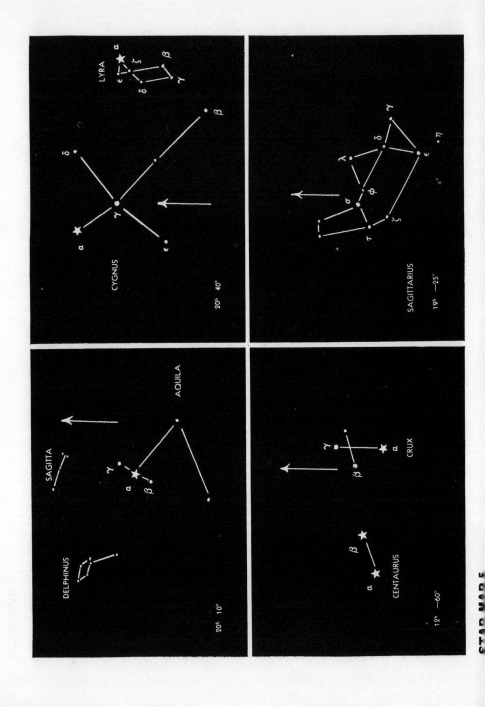

STAR MAPS

THE SKY

through a lens, a flashlight supplying the illumination. Not only is it possible to view the map through the "Stellarscope" so as not to obstruct the vision of the other eye, but the lens is so placed that the image appears projected on the sky when both eyes are kept open. The scale is such that the map image may be superimposed upon the actual sky and positive identification obtained. The constellations included with the device are the ones most easily identified, but additional maps will be available for a more intensive and extensive study.

IN ORDER to prevent students in the middle northern latitudes from wasting time in looking for constellations that are not visible, we have constructed Charts 3 and 4. Chart 3 gives the names of the constellations in their relative positions for those between the celestial equator and the north celestial pole, while Chart 4 is for the southern hemisphere. Certain constellations in both charts are visible in the middle northern latitudes. Arcs of circles marked horizon 0, II, IV, etc., indicate the position of the horizon for Lat. 40° N. at 0, II, IV, etc., hours of sidereal time. Any one of these arcs divides the circular chart into two unequal parts, the larger portion always containing the celestial pole marked "N.P." or "S.P."

To find the constellations visible in Lat. 40° N. at any time, first compute the sidereal time, then select the horizon with sidereal time nearest that computed. The constellations visible in Chart 3 will be those in the larger segment bounded by this horizon and the equator, while those visible in Chart 4 will be found in the smaller portion. Therefore, according to Chart 3, at 0 hr. sidereal time Aquila, Sagitta, Lyra, etc., will be visible in Lat. 40° N., and Virgo and Leo invisible; according to Chart 4, Cetus and Aquarius will also be above the horizon.

The charts may also be used to find the general direction of the constellation. The middle point of the horizon circle in Chart 3 is always the north point, in Chart 4 the south point, while the intersections of the horizon arc with the equatorial circle are due east and due west.

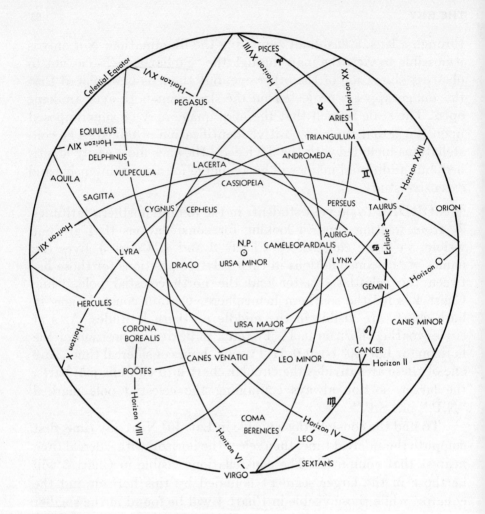

CHART 3

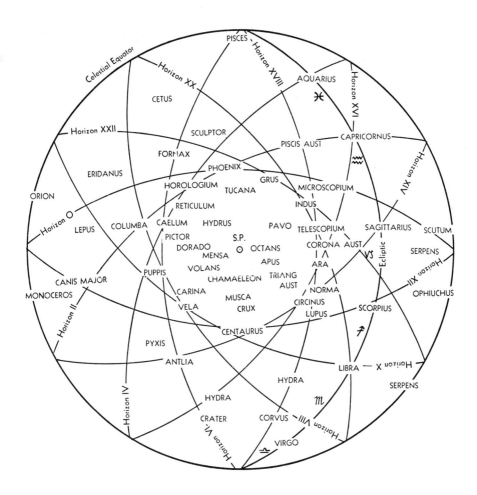

Southern Hemisphere

CHART 4

It may be verified by Chart 3 that the constellations Cassiopeia, Cephus, Draco, Ursa Minor, Camelopardalis, and most of Ursa Major are always above the horizon of Lat. 40° N. These constitute the circumpolar constellations; and, though they contain no first-magnitude stars, many of the second and higher magnitudes can be identified because of their grouping.

In Ursa Major is found that well-known cluster of seven stars, commonly termed the "Dipper," six of the stars being of about the second-magnitude and one (δ) of the fourth. Their names, α Ursae Majoris, etc., are according to their order and are easily remembered, provided the student familiarizes himself with the Greek alphabet. For those interested in names, we note that α, β, γ, δ, ε, ζ, and η Ursae Majoris are, respectively, Dubhe, Merak, Phecda, Megrez, Alioth, Mizar, and Benetnasch. In England this group is known as "Charles's Wain" because of a fancied resemblance to a wagon drawn by three horses; but it is only a part of the Greater Bear, the entire constellation containing one hundred and thirty-three stars visible to the eye. Inasmuch as there is little resemblance to a bear, the tail of this imaginary creature being the handle of the Dipper, it is worthy of remark that both the shepherds of Chaldea in Asia and the Iroquois Indians of America gave to it this name of "Greater Bear." We find a variety of conflicting stories concerning this group of stars. Most of the Greek myths agree in that originally this bear was the nymph Callisto and, through misadventures caused by the jealousy of Hera (Juno), both she and her son Arcas were transformed into Ursa Major and Ursa Minor. According to the star lore of the Pawnee Indians, the bowl of the Dipper was a sick man on a stretcher followed by the medicine man (ε) and his wife (ζ). Now, near Mizar (ζ), is a faint star of fourth magnitude called Alcor by the Arabs, said to mean "The Test," though anyone with fair eyesight can easily discern it. In the Pawnee Indians' legend, this faint object became the medicine-man's wife's dog.

We hurriedly interrupt our description of the stars with a warning to the student totally unacquainted with the skies—he must not feel obliged to recognize and remember in a single night *all* the objects here mentioned. The stars will remain practically unchanged in orientation; so there really is no need for haste. Though we urge him to read this section completely, we also recommend that he commence with those constellations which are visible at a convenient hour and for which maps are included. After he is confident of his ability to locate them, he should enlarge his knowledge, using as reference the additional material offered. In what follows, we shall limit ourselves to naked-eye observations. The telescopic appear-

THE SKY

ance and physical properties of these objects are reserved for the last chapter.

The Dipper provides us with convenient examples of angular distances in the sky. Dubhe (α) and Merak (β) are about 5° apart; Dubhe and Benetnasch (η), about 25°; while Alcor ("The Test") and Mizar (ζ) have a separation of a little more than 10'. As it is always above the horizon in the middle northern latitudes, the Dipper is a good starting-point for locating other objects. Merak (β) and Dubhe (α) are termed the POINTERS, for a line from the former to the latter, when extended five times the distance between the pair (30°), terminates close to the pole star Polaris (α Ursae Minoris).

For unaided eye observations, we may take the pole star as situated at the north celestial pole, though at present it is a little more than a degree away from the point where the earth's axis, extended, pierces the celestial sphere. We have already mentioned that the position of the earth's axis, as well as that of the equator, is slowly changing, so that the celestial pole describes a circle among the stars every 25,800 yr. This motion—*precession*—will tend to make our pole star a truer pole star by bringing the north celestial pole closer to Polaris. In 2120 A.D. the celestial pole and Polaris will be only $\frac{1}{2}$° apart; but thereafter the distance will increase to such an extent that, a few thousand years later, we will be compelled to seek another pole star in the constellation of Cepheus. On the other hand, four thousand years ago the celestial pole was near the star Thuban (α Draconis, right ascension 14^h2^m, declination 64°.8, magnitude 3.6). The "pointers" for the pyramid-builders might have been Phecda (γ) and Megrez (δ) in Ursa Major, for, by drawing a line from γ to δ and extending it four times their distance apart, we reach a point fairly close to Thuban.

Polaris (α) and Kochab (β) are the only second-magnitude stars in Ursa Minor; and, by joining these two stars with the fainter members of the group, the Little Dipper is formed. Draco ("Dragon") is represented in older maps as a long, coiled serpent with its tail-end separating the two bears and its body almost completely circling the littler bear, while the four faint stars which form its head are close to the foot of Hercules. The only second-magnitude star in Draco is in this head, Etanin (γ Draconis, right ascension 17^h54^m, declination 51°.5). Camelopardalis ("Giraffe") is also adjacent to Ursa Major and Ursa Minor, but it contains only such faint stars that it need not concern the beginner. The same applies to Lynx.

A line drawn from Megrez (δ Ursae Majoris) through Polaris and continued an equal distance will strike Caph (β) in Cassiopeia. The stars in this group are easily recognizable by the W they form, Caph (β Cassiopeiae) finishing the last stroke. This second-magnitude star has a right ascension of practically 0 hr.; consequently, when it is on the meridian, the vernal equinox is also there and sidereal time is 0 hr. We might regard Polaris and Caph as forming the hour hand of a 24-hr. sidereal clock, with imaginary numbers from 0 to 23 running *counterclockwise* about the pole, 0 being placed at the zenith and 12 at the north point. According to legend, this W is the throne of Queen Cassiopeia of Ethiopia. Because she boasted that her daughter, Andromeda, surpassed the Nereides or sea nymphs in beauty, a sea monster, Cetus, was sent to devastate the land. To appease the angry deities, King Cepheus was prevailed upon to bind his daughter to a rock, to be devoured by the monster. We remember how Perseus, the future son-in-law, came to the rescue. All the principal actors in the drama are to be found in the sky; even Perseus' trophy of a former combat, the Head of Medusa, may be traced, as well as Pegasus, the winged horse that sprang from Medusa at her demise.

Cepheus is situated between Cassiopeia and Draco and contains no second-magnitude stars. If the last stroke of the W of Cassiopeia, that is, the line segment joining α to β, is extended about three times, the terminus will be close to Alderaimin (α Cephei, magnitude 2.7). The other conspicuous star in this group is β Cephei, lying in the line joining Alderaimin to Polaris, being one-third this distance from the first named.

By extending the last stroke of the W in the opposite direction, from β to α, a little more than four times, Almach (γ Andromedae), a second-magnitude star, is located. The other two second-magnitude stars in Andromeda—Mirach Mizar (β) and Alpheratz (α)—are roughly equally spaced, with Almach on a line that runs parallel to the W of Cassiopeia. Furthermore, Alpheratz is easily identified by the huge square, 15° on a side, which it forms with three of the brighter stars (second- to third-magnitude) in Pegasus, namely, Scheat (β Pegasi), Markab (α), and Algenib (γ). As a further check, we note that Alpheratz (α Andromedae) and Algenib (γ Pegasi) have right ascensions near 0 hr., and hence are close to the arc of a great circle from the pole through Caph (β Cassiopeiae). This arc, when extended 15° beyond Algenib, locates the vernal equinox in the constellation of Pisces. The constellation Andromeda receives most of its fame from a hazy patch of light, located on a line through Mirach Mizar toward Cassiopeia, at right angles to another line connecting Mirach Mizar (β) and Almach (γ), its distance from β being about one-half the distance between these two stars. This patch is barely visible in a clear moonless sky; but it constitutes what is called the Great Nebula in Andromeda, an aggregate of hundreds of millions of suns.

By continuing the line of stars in Andromeda in the opposite direction to the Square of Pegasus about a half again as far, we locate the Segment of Perseus. It consists of several stars in a line curving toward Ursa Major and contains a second-magnitude star, Mirfack (also called Algenib or α Persei). In the midst of a group of small stars which mark the Head of Medusa, and forming a right triangle with Almach (γ Andromedae) and Mirfack, is Algol (β Persei). Its name, which means "The Demon," seems fitting, for it winks—that is, ordinarily it is of magnitude 2.2 but every $2\frac{1}{2}$ days it diminishes rapidly in brightness and descends in a period of $3\frac{1}{2}$ hr. to magnitude 3.4. It then rekindles, and in $3\frac{1}{2}$ hr. more is as brilliant as ever—a phenomenon that everyone can verify.

East of Perseus is Auriga ("The Charioteer" or "The Waggoner"). It is located by the first-magnitude star Capella ("The She-Goat," as it was known in antiquity), the fifth brightest star in the sky. Near this bright yellowish star are three fainter ones called "The Kids."

South of Perseus and Auriga is Taurus ("The Bull") with its brilliant, reddish first-magnitude star Aldebaran in the midst of a group of fainter stars called the Hyades, the brighter ones of which along with Aldebaran form a V, the head of Taurus. The frequent references to the sacred bull in ancient religions may be traced to this constellation. In mythology it was the form assumed by Zeus when he carried off Europa. Only the head, horns, and fore quarters of the Bull are discernible, the explanation being that he is swimming through the ocean.

On the back of the Bull is a little group of six stars of the fourth magnitude and one of the fifth, forming the outline of a tiny dipper, the Pleiades. Many are the allusions to this group of stars; the old Greek myths say that they were the seven daughters of Atlas and Pleione and, in some versions of the legend, made away with themselves from grief at the death of their sisters, the Hyades, both groups being placed in Taurus. Others have it that the Pleiades were pursued by the hunter Orion but that, through the intervention of the gods, they were metamorphosed into doves and placed among the stars. The brightest member of the group is Alcyone (η Tauri). There are also many references to a lost sister, namely, that the group originally had seven stars and one disappeared, leaving only six. As a matter of fact, merely six stars of the cluster can be plainly seen, but on a very clear night a sharp-eyed observer may find five more, making a total of eleven sisters.

The Pleiades were also known as "The Many Little Ones," "The Hen and the Chickens," and "The Little Eyes." Several ancient calendars began their year when the Pleiades crossed the meridian at midnight, and it is said that petitions were always granted in ancient Persia on this date. Their sister-group the Hyades, or "The Rainers," wept so bitterly over the death of their brother Hyas that Jupiter, touched by their grief, changed them into stars. In ancient times, their rising simultaneously with the sun usually announced rainy weather.

In Chart 3 it will be noticed that the ecliptic passes through Taurus. Constellations which lie along this apparent path of the sun among the stars are called ZODIACAL CONSTELLATIONS, the zodiac being a belt of the sky extending 8° on either side of the ecliptic. The word itself is derived from the Greek *zōdiakos*, meaning "of animals"; and, from the earliest recorded times, the twelve constellations occurring on the ecliptic, with the exception of Libra ("The

Balance"), were the names of living creatures. The zodiac forms a sort of street in the sky, down the center of which moves the sun, while the moon and the planets, in their more apparently erratic paths, never venture beyond its borders.

The zodiac is divided into twelve sections, each of 30°, called the SIGNS OF THE ZODIAC. They are named after the twelve zodiacal constellations; and in many almanacs the position of sun, moon, and planets is indicated by the sign in which the object happens to be. On or about March 21 the sun is at the vernal equinox, or, as it is also frequently called, the "first point of Aries." In this scheme, the sun is said to "enter Aries." A month later, the sun enters Taurus, the different signs of the zodiac being, in order,

Spring signs
1. ♈ Aries ("The Ram")
2. ♉ Taurus ("The Bull")
3. ♊ Gemini ("The Twins")

Summer signs
4. ♋ Cancer ("The Crab")
5. ♌ Leo ("The Lion")
6. ♍ Virgo ("The Virgin")

Autumn signs
7. ♎ Libra ("The Scales")
8. ♏ Scorpius ("The Scorpion")
9. ♐ Sagittarius ("The Archer")

Winter signs
10. ♑ Capricornus ("The He-Goat")
11. ♒ Aquarius ("The Water-Bearer")
12. ♓ Pisces ("The Fishes")

On referring to Chart 3, it is seen that the points where the ecliptic crosses the celestial equator are in Pisces and Virgo, the vernal equinox being in the first and the autumnal equinox in the second. This seems strangely inconsistent with the signs, but the explanation is in the motion of the equinoxes. The sun's path among the stars is the same as it was tens of thousands of years ago; but about every two thousand years the vernal equinox moves 30° westward along the ecliptic with respect to the stars, i.e., an angular distance equivalent to one sign. Instead of changing the seasonal signs with this motion, the signs are still partitioned among the seasons in a manner corresponding to the sun's position in the constellations twenty-one centuries ago. When Hipparchus located the first point of Aries, he found the vernal equinox in Aries, not in Pisces, as it is now. Today these signs do not represent constellations; they may be regarded as a method of indicating angular distance from vernal equinox along ecliptic—a quantity which

THE SKY

is more accurately and simply expressed in degrees. Were we to adopt signs corresponding to present position, we should rename the Tropic of Cancer, calling it the Tropic of Gemini, for it is in the Twins that the sun now stands when it is directly over Lat. 23°27′ N.; similarly, the Tropic of Capricorn now belongs to the Archer.

It has been suggested that these signs owe their names to phenomena associated with seasonal changes. Thus, Aries ("The Ram") was so called because it rose with the sun in the spring and the Chaldean shepherds regarded the flocks as their most prized possessions. Taurus represents the herds, which were esteemed next in value; and Gemini, the bringing-forth of young by the creatures of the field. When the sun is in Leo, the brooks are dry and the lion leaves his lair and becomes a terror in the land. (Another interpretation is that Leo represents the strength with which the sun beats down during this month.) Virgo is for the maidens who glean in the fields after the summer harvest, while Libra ("The Scales") very aptly indicates the equality of day and night at the autumnal equinox. Vegetation decays as the sun continues on its southward journey; therefore the Scorpion, which stings as it recedes and brings death to mankind, is in evidence. Sagittarius is the hunting month, while Capricornus ("The He-Goat"), which delights in climbing, indicates that the sun's altitude is starting to increase, Aquarius denotes the rainy season, and Pisces the month for fishing.

This explanation of the zodiacal names has one difficulty. Three thousand years before Christ, the vernal equinox was in Taurus and not in Aries, and it is a question if this was the original interpretation. It seems quite likely that Libra was formed at a later date from the claws of the Scorpion. Perhaps, to begin with, this association of season and sign was based on the constellation that *preceded* the rising of the sun (incidentally, the one *containing* the sun would be invisible); and, if this guess is correct, then the doubt is removed.

The only zodiacal constellations possessing first-magnitude stars are Taurus, Gemini, Leo, Virgo, and Scorpius. Gemini is adjacent to Taurus and Auriga, and its principal stars are Pollux (β Geminorum) and Castor (α). Both stars are of about the same brilliancy, though Pollux, the more southerly of the pair, is a trifle brighter and falls in the first-magnitude class, while Castor is in the second; this and their separation of a little less than 5° serves to identify them. In mythology, Castor was a celebrated horse-trainer, Pollux a boxer, and they were inseparable in their adventures. Sailors regarded them as their patron deities—St. Paul sailed for Italy in a ship whose sign was Castor and Pollux (Acts 28:11)—and today we find them in the *Nautical Almanac* among the other stars used in navigation. Gemini is distinguished by two nearly parallel rows of stars running southwest from Castor and Pollux. The row starting from Pollux and terminating with Alhena (γ Geminorum, the only other second-magnitude star in the constellation besides Castor), when continued half again as far, points to the ruddy, first-magnitude star Betelgeuse (α Orionis).

Orion, "The Warrior" or "The Belted Giant," is conspicuous in that it has two first-magnitude stars—the white star Rigel (β Orionis) and the red star Betelgeuse. This pair are about 20° apart and are separated by a row of three second-magnitude stars 1½° apart, known as the Belt of Orion, the Band of Orion, Jacob's Rod, or the Yardstick. As it is situated on the celestial equator, the Belt rises at the east point and sets at the west. Running southward from the Belt are three fainter stars, the Sword of Orion, the central one of which appears a trifle hazy. Actually, this "fuzzy" spot in the sky is a vast volume of luminous gas—the Great Nebula of Orion.

Using a little imagination, we picture Orion in the sky as a mighty giant who, with a club in his right hand and a lion skin in his left, faces the charging Bull. Betelgeuse (α) and Bellatrix (γ) mark his right and left shoulders, respectively; Rigel (β) is in the left foot and Saiph (κ) in the right knee. In strange contrast we find just south of Orion, crouching at his right foot, four stars forming the timid Lepus ("The Hare").

Southeast of Orion and following him across the sky is Canis Major ("The Greater Dog"), containing the brilliant star Sirius. This was the watchdog of the Egyptians, for, though it could not bark, they knew that the Nile would soon overflow its banks when they saw Sirius' brilliant light in the early morning just before dawn. Precession, however, has moved the summer solstice from Cancer to Gemini, and no longer is Sirius the watchdog of the Nile; only the ruins of temples remain to indicate its rising point in ages past. The orientation of temples directed toward the rising point of the sun at the time of the summer solstice is still correct, for that point remains essentially unchanged through the ages. Those directed to Sirius, however, were not so permanent, for precession slowly changed its rising-point and every few centuries a new temple was erected.

The Greeks and Romans associated Sirius, the Dog star, with the heat of summer and said that it burned up the fields and killed the bees; hence the expression "dog days" for the hottest period of the year. The Lesser Dog, Canis Minor, is found just south of the Twins; and its first-magnitude star, Procyon (α) was also considered an object of evil omen, the harbinger of storms. Bearing the same relative position to Procyon as Castor does to Pollux is a third-magnitude star Gomeisa (β Canis Minoris).

Cancer, the zodiacal constellation east of Gemini, contains no bright stars; but in its center is a faint luminous spot called Praesepe ("The Beehive"), which ordinary binoculars resolve into many stars, a large telescope into thousands.

Continuing along the ecliptic, we pass from Cancer to Leo ("The Rampant Lion"), one of the most spectacular constellations in the zodiac. The principal stars are arranged in the form of a sickle or reversed question mark, and the first-magnitude star Regulus (α Leonis) is almost exactly on the ecliptic. Zosma (δ) lies in the back of the lion; Chort (θ), in the hind quarters; while Denebola (β), the second brightest star in the group, is in its tufted tail.

THE SKY

A trifle northeast of Leo is Coma Berenices ("Berenice's Hair"), a cluster of faint stars. Berenice was a queen of Egypt; and the story goes that, being worried about her husband's return from a dangerous expedition, she promised to consecrate her crowning glory to the gods if he should return safely. Soon after the fulfilment of her vow, the lovely tresses disappeared from the temple. Fortunately for all parties concerned, an astronomer announced that they had been transferred to the heavens, in proof of which he pointed out this hitherto un-named cluster of stars. There is a sequel to the legend. Originally, Leo's tail terminated in this region of the sky; but, to make room for the queen's hair, he twisted it so that it now ends in Denebola.

North of Coma Berenices is Canes Venatici, "The Hunting Dogs." The constellation contains only one bright star of third-magnitude, Cor Caroli (α), which is found on a line joining Benetnasch (η Ursae Majoris) to Denebola (β Leonis). Cor Caroli and Berenice's Hair divide this line into three equal parts.

Virgo is south of Coma Berenices and southeast of Leo, being represented by a maiden with folded wings, bearing in her left hand an ear of corn. She is said to have been the last of the immortals to bid the earth farewell. The first-magnitude star Spica (α Virginis) is in the ear of corn. Five third-magnitude stars—ϵ, δ, γ, η, and β—make a corner known to the Arabian astronomers as "The Retreat of the Howling Dog."

Northeast of Virgo is Boötes ("The Bear-Driver") which is easily located through its magnificent first-magnitude star Arcturus (α Boötis), the light of which was used to give the initial impulse for the illumination of the Century of Progress Exposition at Chicago, 1933. To find Arcturus without the aid of coordinates, merely continue the handle of the Dipper, bending it a trifle. Boötes is represented as a hunter grasping a club in his right hand, while with his left he holds in leash his two greyhounds (Canes Venatici), with which he pursues the Great Bear around the pole star.

Bordering Boötes is Corona Borealis ("The Northern Crown"), a semicircle of stars of which the brightest, Alphacca (α Coronae Borealis), is of second magnitude. This semicircle of stars is turned to the northeast away from Arcturus and toward Hercules ("The Kneeling Hero").

The hero Hercules, because of his twelve labors, is famous in mythology; but, nevertheless, this huge constellation contains no stars brighter than third magnitude. Its situation between the first-magnitude star Vega of the constellation of Lyra ("The Harp") and the Northern Crown makes it easy to find; and the student, after discovering the group, may construct his own figures out of the third- and fourth-magnitude stars therein—a butterfly, for example.

Positive identification of the blueish-white star Vega (α Lyrae), the second brightest star visible in the middle-northern latitudes, is obtained through the presence of two fourth-magnitude stars ϵ Lyrae and ζ Lyrae which, with Vega, form an equilateral triangle, sides about 2° in length. Furthermore, ζ Lyrae forms a small parallelogram with three other stars. Twelve thousand years from now, the inhabitants of the globe will have Vega for their pole star, a rôle it played fourteen thousand years ago. How fortunate they will be to have so bright a star to mark the north, but we may console ourselves with the thought that Vega will never be as close to the north celestial pole as Polaris is now.

People possessing good vision should look closely at ϵ Lyrae, for keen eyes will discern that it is in reality two stars, though their separation is only $3\rlap{.}'5$. Since this *double* star is mentioned neither by the Greeks nor the Arabs, the statement has been made that modern eyesight excels that of the ancients. Another interesting object is β Lyrae at one vertex of the parallelogram. As was true in the case of Algol ("The Demon"), β Lyrae varies periodically in brightness. Its period is 12 days and 22 hr., and it fluctuates by 1 mag.

East of Lyra and northwest of Pegasus is Cygnus ("The Swan"), the principal stars of which form the Northern Cross. The first-magnitude star Deneb (α Cygni) is at the top of the Cross; Sadr (γ), of second magnitude, at the center; while three third-magnitude stars mark the remaining arms.

South of the Cross on the celestial equator is Aquila ("The Eagle"), which contains the first-magnitude yellowish star Altair (α Aquilae), easily recognizable by its position as the central star in a row of three. Between the Eagle and the Swan is a compact group of faint stars, Sagitta—the Arrow that has missed both birds. Northeast of Aquila is another diamond-shaped group known as "The Dolphin" or, sometimes, "Job's Coffin."

Perpetuating the memory of Aesculapius, the Father of Medicine, is Ophiuchus, or the "Serpent-Bearer," situated south of Hercules and west of Aquila on the celestial equator. Because of his skill in restoring the dead to life, Pluto, who was the god of the underworld, protested; and Jupiter terminated the physician's embarrassing career with a thunderbolt. Serpents were sacred to Aesculapius, for he believed that, in renewing their skin, they also renewed their youth; and in the sky he grasps an enormous writhing serpent, the ends of which, Serpens Caput and Serpens Cauda, constitute two other constellations. Ras Alhague (α Ophiuchi) and Cebalrai (β) form a pair 5° apart, similar to the Twins, except that they are about half as bright.

Bordering the southern extremity of Ophiuchus are three zodiacal constellations—Libra, Scorpius, and Sagittarius. Of these, Scorpius is the most easily recognized because of its fiery-red first-magnitude star Antares (α Scorpii), which marks the Heart of the Scorpion. A triangular group of second-, third-, and fourth-magnitude stars mark the head, in which Acrab (β Scorpii) and Dschubba (δ) are the most conspicuous. A good way to remember this group is to picture the triangle as a kite with a long tail—the Tail of the Scorpion—which starts at Antares, extends southeastward, and then curves gracefully northward. According to legend, it was the Scorpion which stung Orion in the heel after that mighty hunter had boasted that no animal could kill him. Obviously, Orion has learned

THE SKY

his lesson for in the middle-northern latitudes the two constellations are never seen above the horizon simultaneously, Orion being prominent in the winter skies, Scorpius in the summer.

To the eye a quadrilateral of four faint stars, Libra is located just northwest of the Scorpion's head. Commencing at the boundaries of Libra and Scorpius is the tail of Hydra, a long, straggling serpent. Its head is about 10° from Procyon (α Canis Minoris), so that Hydra extends for more than 100°. Its principal star is Cor Hydrae, a lone star of the second magnitude, which the student may locate through its co-ordinates (right ascension 9^h23^m, declination $-8°\!.2$) or by remembering that it is 23° south and a trifle west of Regulus (α Leonis), forming with this star and Procyon (α Canis Minoris) a right triangle with itself, Cor Hydrae being at the vertex. On the back of Hydra we find Corvus ("The Crow"—four third-magnitude stars in the form of a quadrilateral and 15° southwest of Spica), along with Crater ("The Cup"—a semicircle of six fourth-magnitude stars 15° west of Corvus).

The zodiacal constellation Sagittarius ("The Archer") follows the Scorpion and is represented as a centaur with his bow bent as if about to shoot an arrow at that poisonous creature. The stars of the Archer may be joined together so as to form a teapot; and, as the Scorpion travels across the sky followed by this teapot, the latter becomes tilted, as though an unseen hand were pouring hot tea on the tail of the Scorpion. The second-magnitude stars in this group are Kaus Australis (ϵ Sagittarii) in the bow and σ Sagittarii, the last being part of the bowl of the Milk Dipper (formed by joining λ, ϕ, σ, τ, and ζ).

Capricornus and Aquarius contain no conspicuous stars; but just south is Pisces Austrinis, with the first-magnitude star Fomalhaut. Pisces, situated south of Andromeda and Pegasus, can only be traced on a clear, moonless night.

Aries completes our circle and brings us back to Taurus. Its most noted star is Arietis (alpha of Aries) about 20° south of Almach (γ Andromedae). Midway between Arietis and Almach is a figure of three stars—Triangulum. Drawing a line from Almach to Arietis and continuing it one and one-third times the distance between the pair, we arrive at Mira (o Ceti, right ascension 2^h14^m declination $-3°\!.4$). This star of Cetus justly lives up to its name, "The Wonderful." Ordinarily, it is of the second magnitude for about 15 days, when it decreases for 3 mos. until it becomes invisible to the naked eye. This period of darkness lasts almost 6 mos., and then it rebrightens in 3 mos. to its original magnitude. Occasionally, however, it fails to intensify according to schedule, and once it actually became brighter than Aldebaran.

Though we have by no means given an exhaustive list of the stars visible in middle-northern latitudes, we shall terminate this section with a word or two concerning the appearance of the sky for an observer residing in the middle-southern latitudes. The constellations here are, in a sense, reversed—Orion stands on his head, presumably juggling Canis Major on his feet; all our circumpolar constellations are hidden from view; due south, midway between the horizon and the zenith, is the south celestial pole. Such constellation names as "Telescopium" and "Microscopium" indicate that this region of the sky was charted much later. No conspicuous star marks the south celestial pole; Sigma of the constellation Octans is about 1° away, but it is of sixth magnitude and barely visible to the eye.

To find the direction south, we locate Crux, more commonly called the Southern Cross; but we must be careful not to err and select the larger and fainter False Cross of Argo ("The Ship"). Fortunately, there are the pointers, α and β Centauri (both of first magnitude) which direct us to the top of the Cross. We join the top and bottom (γ Crucis and Acrux) and continue the line five times to locate the south celestial pole, verifying our result by noting whether this point is roughly equidistant from Achernar (α Eridani) and the brilliant Canopus (α Carinae).

EXTENDING completely around the entire sky is a hazy band of light—the Milky Way. Its width averages about 20°; and its middle forms a great circle of its own in the heavens, cutting both equator and ecliptic. One way of specifying the position of the Milky Way is by the co-ordinates of a point on its northern side 90° from its central circle—the north pole of the Milky Way—right ascension $12^h 40^m$, declination 28°; i.e., at 12:40 sidereal time in latitude 28° the north pole of the Milky Way is overhead while the Milky Way itself is right on the horizon. We may also follow its path through the constellations (see Charts 3 and 4).

We locate it in Cepheus and Cassopeia, where it attains its maximum northern declination and is about 20° wide, tracing its irregular outline through Perseus, Auriga, Taurus, and Orion, as it narrows down to 5°. It widens as it crosses the equator in Monoceros to cut the northeast corner of Canis Major. In the southern hemisphere it is seen to spread and become divided by dark lanes, as it continues through Puppis, Pyxis, and Vela, to reach its maximum southerly declination in Crux and Musca. There, between Crux and the pointers (α and β Centauri), we find the Coal Sack—a dark oval in the midst of the Milky Way. Heading northeastward, it passes through Circinus, Lupus, and Norma, becoming visible once more to northerners through Scorpius and Sagittarius. As it enters Ophiuchus, the Milky Way seems divided; and this bifurcation continues through Ophiuchus, Serpens, Aquila, Sagitta, and Cygnus.

To the eye, it appears as a faint, luminous band of light; but a telescope resolves it into hundreds of thousands of stars. In Grecian myths it is the dust

THE SKY

stirred up by Perseus in his haste to rescue Andromeda; in astronomy, the Milky Way marks the plane of our galaxy, an aggregate of billions of stars, of which our sun is only one.

THE prime purpose of this chapter was to acquaint the student with the sky, and this is our excuse for the many charts and tables. We hope that they will enable him to locate celestial objects for any date, time, or latitude during his lifetime; but he need not limit himself to this comparatively short interval of time. With Chart 2 it is possible to extend the period ten or twenty thousand years.

We have repeatedly mentioned precession (the motion of the earth's axis due to the attraction of the sun and moon on the equatorial bulge of the earth) and its alteration of the positions of the stars. To obtain a simple picture of the precessional motions of the stars, we mentally transport ourselves to the arctic circle. When our sidereal time equals 18 hr., the ecliptic (the apparent path of the sun among the stars) will coincide with our horizon—a phenomenon observable only at Lat. $66°.5$ N. or $66°.5$ S. At that instant the vernal equinox is due east; the autumnal equinox, due west; and at the arctic circle the NORTH POLE OF THE ECLIPTIC (a point 90° from the ecliptic) is directly overhead. Now at 18 hr. sidereal time we stop the diurnal motion of the earth, for it would be much too difficult to follow the rotation of the earth through the 25,800-yr. precessional cycle. When we have occasion to start the earth in its daily motion once more, we must be careful that we spin it about its present axis; and we therefore note the place where the earth's axis, extended, pierces the celestial sphere—altitude $66°.5$, azimuth 0°. As long as we remain in our present position on the arctic circle, the north celestial pole will have these co-ordinates; and, to avoid any possible doubt, we permanently imbed a long pole in the ground, inclining it to the north celestial pole.

As we revolve about the sun on a non-rotating earth, the sun moves annually around our horizon (the ecliptic). Let us extinguish the sun, so that we may watch the stars. At first glance they seem stationary; but, as the centuries go by, they appear to move slowly clockwise about our zenith (the pole of the ecliptic), completing their circuit in 25,800 yr. To express it in different words, the stars appear to travel parallel to our horizon (the ecliptic), their altitude (which is also their distance from the ecliptic) remaining unchanged, while their azimuth changes by 14° every 1,000 yr. The constellations on our horizon (the ecliptic)—the zodiacal constellations—remain always on this circle. But in the one hundred and forty-fifth century our inclined pole is pointing to Vega, whereas in the twentieth century it pointed to Polaris.

Relative to us the north celestial pole, the ecliptic (our horizon) and the celestial equator seem fixed in the sky, while the stars appear to revolve about a vertical axis. Actually, it is the earth's axis, the direction of which we defined by the inclined pole, which is describing a cone about our vertical (the line directed to the pole of the ecliptic).

We have remarked that the ecliptic (our horizon) and the celestial equator do not change their position relative to us, and therefore their mutual inclination remains unchanged. Inasmuch as the seasons are due primarily to the angle between these two circles, precession does not in itself affect the seasons.

These imaginary observations we repeat with Chart 2, thereby finding the manner in which the co-ordinates of any particular star change with the time. Using the special scales T_1, T_2, C, and R_1, we find the altitude and azimuth of the star at the arctic circle for sidereal time 18 hr.

T_1 (hour angle) $+C$ (declination)$\rightarrow R_1$ (altitude),
R_1 (declination)$+C$ (altitude) $\rightarrow T_2$ (azimuth),

for the twentieth century. For every thousand years in the future, we change the azimuth by 14°, moving the star clockwise about the zenith. In other words, if the azimuth is east, we *subtract* for future positions 1.395° times the number of centuries that will elapse; if the azimuth is west, we *add* this quantity. (*Note:* This may carry the star past the south or north points, i.e., azimuth 200° W. should be replaced by azimuth 360−200 =160° E.; azimuth −10° E. becomes azimuth 10° W.) Past positions may likewise be derived; if azimuth is east, *add* 1.395° times the number of centuries which have elapsed; if azimuth is west, *subtract* this product.

THE SKY

With the derived azimuth, the original altitude and the known sidereal time (18 hr.), we find the corresponding declination and right ascension of the star.

T_2 (azimuth) $+ C$ (altitude) $\rightarrow R_1$ (declination) ;
R_1 (altitude) $+ C$ (declination) $\rightarrow T_1$ (hour angle).

With these new co-ordinates the altitude and azimuth are obtained for *any* latitude, time, and date in precisely the manner described for the twentieth century. By "date" is meant that which is indicated by a calendar in which the vernal equinox occurs on or about March 21. For periods of less than 10,000 yr., the Gregorian calendar will be sufficiently accurate. For longer intervals, our calendar is the Gregorian with the leap-year rule modified so as to correct for the excess of a day in 3,200 yr. With this meaning for "date," Table 5 for the sun's declination remains essentially unaltered and may be used for any epoch.

As an example, we shall find the right ascension and declination of Acrux (alpha of the Southern Cross) for the one hundred and fiftieth century (14,900 A.D.). We derive for Acrux in the twentieth century:

Right ascension	12^h22^m	(Table 3)
Declination	$-62°6$	(Table 3)
Hour angle at 18:00 sidereal time	5:38 W.	
Altitude at arctic circle	$-53°$	(Chart 2)
Azimuth at arctic circle	130° W.	(Chart 2)

For the one hundred and fiftieth century we add $1.395 \times 130 = 181°$ to azimuth, since it is west. At the arctic circle, 18 hr. sidereal time, in the one hundred and fiftieth century:

Azimuth	(311° W. *or*) 49° E.
Altitude	$-53°$
Declination	$-35°$
Hour angle	10^h00^m E.
Right ascension	$18+10=28$ *or* 4^h E.

These equatorial co-ordinates, right ascension 4 hr., declination $-35°$, for Acrux in the one hundred and fiftieth century apply to any latitude and enable us to determine the star's position anywhere on earth. For example, at the beginning of winter (December 22) in the year 14,900 A.D., Acrux will cross the meridian at 10 hr. past noon and in latitude 40° N. will have a maximum altitude of 15°.

Starting with the twentieth-century co-ordinates of Thuban (α Draconis), right ascension 14^h2^m, declination $64°8$, we follow this scheme and discover that this particular star had a declination near 90° (i.e., the north celestial pole) in 2900 B.C., the estimated date of the building of the Great Pyramid. The student may also verify that the year 400 B.C. was about the last time the Southern Cross was visible in Lat. 40° N.

IF THE maps included with this material are used to find the orientation of the constellation in the sky after finding its direction, it should be remembered that the arrow on the map indicates the direction of the north celestial pole for the twentieth century, a point near the star Polaris. For hypothetical positions in the past or future, the arrow is to be directed toward Polaris, the altered position of which is obtainable by the method just outlined from its twentieth-century co-ordinates (right ascension 1^h22^m, declination $88°.8$).

With the charts, the student may solve problems in stellar position (including the sun) for an interval of about one hundred thousand years. We do not advise that he venture further than this period for, even in fifty thousand years, the motion of the stars relative to each other becomes noticeable—the handle of the Dipper is bent and the bowl elongated, while the outline of the chair of Cassiopeia is destroyed.

CHAPTER 4

> Baby moon, 'tis time for bed,
>   Owlet leaves his nest now;
> Hide your little hornèd head
>   In the twilight west now;
> When you're old and round and bright
> You shall stay and shine all night.
> —"Nursery Rhyme"

THE MOON

THIS is a remarkable poem, inasmuch as its description of the lunar phases is consistent with astronomical observations—a by no means common occurrence. How often have we read something like this: "As the hands of the clock on the old church steeple pointed to the witching hour, the full moon, like a golden disk, rose over the silent waters; and Wilbur knew that his love would soon arrive"—and we concluded that either the clock was sadly in need of repairs or the hero had suddenly been transported to the frozen north and was about to propose to a Laplander.

But liberties taken with the moon are not confined to literature alone. We have all encountered drawings of the crescent moon and somewhat grudgingly admired the courage of the artist in placing a fiery star midway between the tips of the horns! Furthermore, we fear the astronomer will never appreciate the artistic merits of directing the horns of the moon toward the sun. If the artist declines this invitation to read on into the causes of the phases of the moon, we beg him at least to look at the object itself in the sky before he places on his canvas such astronomical atrocities as the autumn sunset we have carefully reproduced. For the benefit of those readers who fail to appreciate the astronomers' protests, we shall consider the phases of the moon.

Without doubt you have all heard that the moon is the earth's only satellite, that it revolves about the earth, and that the earth and moon together revolve about the sun. As we watch the moon on its monthly journey among the stars, passing from time to time in front of the stars of the zodiac and describing its approximately circular path in $27\frac{1}{3}$ days, we are observing a real motion. Of course, there may also be some huge boulders, at most a few hundred feet in diameter, that likewise travel about the earth but which have not yet been trapped by telescopic observations.

THE moon's illumination is due wholly to the sun; that is, when we look at the moon, we see light that has traveled from the sun to the moon and then has been reflected back to our eye. Since the moon is spherical in shape, half of it is always in the sunlight and the other half in darkness. For a practical illustration of the phases, hold a white ball before a brilliant light, letting the latter be the sole source of illumination. If the light is not too close to the ball, the ball is divided into a bright and a dark hemisphere. Depending on the position of the ball relative to our eye and the light, we see more or less of the illuminated portion; if the ball is partly between us and the source of illumination, the bright side is facing away from us and we see most of the dark half.

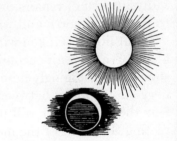

THE MOON

Precisely the same thing happens in the case of the new moon. In the sky the moon is seen as a thin crescent close to the sun, and hence setting soon after the sun. Since the sun is actually four hundred times as far from us as the moon, in this position the moon is partly between us and the sun, and the thin crescent is all we can see of the illuminated half. Some readers may have noticed that the "dark" of the moon seems dimly illuminated, which phenomenon is often referred to as "the old moon in the new moon's arms." The explanation, which we owe to Leonardo da Vinci, is that, when the

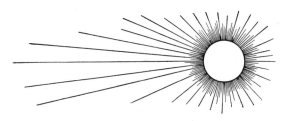

dark half of the moon is facing the earth, the illuminated half of the earth is facing the moon; and, as the earth is reflecting sunlight to the moon, we are seeing "earthshine" on the moon. The bright crescent is due to light traveling from sun to moon to our eye, while the "old moon" is due to sunlight that has traveled from sun to earth to moon and back to our eye.

The moon now waxes; and, when it is a week old, that is, 7 days after it was new, the moon is at right angles to the sun. As before, half of the moon is in sunlight and half in darkness; but, this time, we see 50 per cent of the illuminated half, while 50 per cent of the dark half is directed toward us. The moon is said to be at "first quarter," and it crosses the meridian 6 hr. after the sun.

A week later, practically all of the illuminated disk is visible and the moon is "full," rising with the setting sun and setting with the rising sun. Thereafter the moon wanes, and in another week we again see only half of the illuminated portion. The moon is at "last quarter" and crosses the meridian 6 hr. before the sun. As it continues to wane and approach the sun, the crescent becomes thinner and thinner, the horns always pointing away from the sun. Finally, the moon disappears from sight, to reappear later on the other side of the sun with horns reversed. In passing the sun, the moon may pass either north or south and even occasionally in front of the sun, the condition for a solar eclipse.

Another word of information to the artist desirous of reproducing the crescent or the GIBBOUS (i.e., between quarter and full) moon. The boundary between the dark and light half of the moon is always a circle. Now, if a bowl with a circular rim be drawn, the rim will not necessarily appear as a circle but will probably be an ellipse, the shape depending on the position from which the bowl is viewed. Therefore, when the artist draws the _terminator_ of the moon—this boundary of light and dark—it should appear as half an ellipse. The rules for drawing a moon are: Construct a circle, inside of which place an ellipse just touching the circle at two points. By combining half of the ellipse with half of the circle, either the crescent or the gibbous moon is obtained. If the moon is to be a thin crescent close to the setting sun, then the line joining centers of both sun and moon should be perpendicular to the terminator. The

THE MOON

same rule holds in the sky for any phase, provided we replace this line by a great circle, that is, one with the observer at center. On canvas, however, the circle projects into an ellipse; and we conclude that, to play safe, the artist should make his moon either very young or very old, if he really must have both sun and moon in the picture.

We hesitate to propose rules for determining the relative sizes of the sun and moon. The diameters of both subtend an angle of about one-half a degree in the sky; but, because the moon's distance

from the earth varies appreciably, its *angular* size may be a trifle greater than, equal to, or a trifle less than that of the sun. This, however, is not the important difficulty. There is no doubt that both sun and moon seem considerably larger when they are close to the horizon than when they are high in the sky. We must turn to the psychologist for an explanation of this illusion, for there is no essential variation in their angular size. Perhaps they seem larger when they are near the horizon because this is a better position for comparison with terrestrial objects. But what should the artist do about it? Well, this is not a treatise on art; and, just as we permit a certain amount of poet's license, so must we accord the same courtesy to the painter.

THE length of time from new moon to new moon, or from full moon to full moon, averages $29\frac{1}{2}$ days, which is 2 days greater than the average period of revolution. This is not surprising when we examine the apparent motion of the moon and sun among the stars. While the moon is hastily describing its circuit through the zodiacal constellations, the sun appears to move eastward about 30° so that, after completing one revolution about the earth, the moon must travel another 2 days to overtake the sun. In fact, the number of new moons in a year is one less than the number of revolutions of the moon about the earth.

A study of the moon's path among the stars reveals that it is always close to the ecliptic, its center never deviating by more than 5°9′ from the apparent path of the sun. Actually, it moves approximately in a circle inclined 5°9′ to the ecliptic and intersecting the ecliptic at two diametrically opposite points, the NODES, as they are called. A very careful plotting of the path of the moon from night to night and year to year indicates that this description is hopelessly inadequate. Men have devoted a lifetime trying to find out exactly where the Queen of the Night is going to wander in the starry skies. Mathematical formulae have been derived that fill many volumes, all for the purpose of predicting the future position of the moon.

With respect to the earth, the moon's motion may briefly be described as follows: The moon moves in an ellipse with the earth at one of the foci, obeying the Law of Areas as a first approximation. The attraction of the sun not only tends to retard the moon at new and full moon and increase the period of revolution at the earth's perihelion but also causes the eccentricity of the elliptical path to vary enormously, the inclination of the ellipse to oscillate slightly, the axis of the ellipse to advance, the nodes to move westward along the ecliptic, etc.

AS MOST of us have no intention of devoting a lifetime to the analysis of the capricious behavior of the moon, we shall forget all this and merely remember that the moon is always near the ecliptic, never departing from the latter by more than about 5°. Since the new moon is near the sun, both new and old moons attain their highest position in the sky near the beginning of summer, when the sun also reaches its maximum altitude. This is not so with the full moon, for it is approximately opposite the sun and at its highest when the sun is lowest; thus, for our northern latitudes, the near approach of winter (December 22) is the proper time for moonlight picnic excursions. At the time of the winter solstice the sun is south of the equator by 23°.5; consequently a full moon near this date is the equator plus-or-minus an angle of at most 5°—a result which is more simply expressed as follows: declination = $23°.5 \pm 5°$. Similarly, let us denote by $-23°.5 \pm 5°$ a value between $(-23°.5 - 5° =) -28°.5$ and $(-23.5 + 5° =) -18°.5$, and construct in this notation the accompanying table. For example, on March 20,

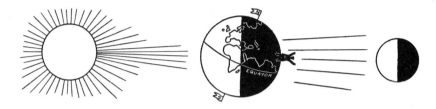

1935, the moon was full; our table informs us that its center was not more than 5° away from the celestial equator.

DECLINATION OF MOON'S CENTER AT DIFFERENT PHASES

Time		New	First Quarter	Full	Last Quarter
Near beginning of	Spring	$\pm 5°$	$23°.5 \pm 5°$	$\pm 5°$	$-23°.5 \pm 5°$
	Summer	$23°.5 \pm 5°$	$\pm 5°$	$-23°.5 \pm 5°$	$\pm 5°$
	Autumn	$\pm 5°$	$-23°.5 \pm 5°$	$\pm 5°$	$23°.5 \pm 5°$
	Winter	$-23°.5 \pm 5°$	$\pm 5°$	$23°.5 \pm 5°$	$\pm 5°$

The curious reader naturally asks what is the maximum angular height of the moon. If the moon traveled in the ecliptic, then its maximum declination would be that of the sun, namely, $23\frac{1}{2}°$, which is the inclination of the equator to the ecliptic. But what of the 5° we mentioned? The answer is contained in the brief description of the moon's motion, which we agreed to forget. Owing to the sun's attraction, the plane of the moon's orbit is slowly turning but always maintaining this inclination of 5° to the ecliptic. It takes about 18.5 yr. for the nodes (those points where the moon crosses the ecliptic) to move completely around; and for one year in this interval each phase of the moon is given an opportunity to attain the maximum declination of $23\frac{1}{2}° + 5° = 28°.5$. Since the maximum altitude of a celestial object is its declination *plus* 90° *minus* the observer's latitude, in Lat. 40° N. the maximum altitude of the moon is $28°.5 + 90° - 40° = 78°.5$. The years in which each phase has

had or will have an exceptionally large yearly variation in declination (from about $-28°$ to $28°$) are:

1930	1949	1967
1931	1950	1968
1932	1951	1969
1933	1952	1970

Perhaps you noticed how very high was the full moon in the winter of 1931 and how low it was during the following summer. If you failed to observe this, look for a repetition of the phenomena in 1950, for, in that year, we shall see the noonday full moon even at Lat. $61\frac{1}{2}°$ N.

ALL admirers of the rising moon realize that it usually rises later on succeeding nights. By "later" we mean with respect to the sun, for our clocks give us corrected sun time. Since the moon is increasing its angular distance from the sun at such a rate that, starting with new moon, after $29\frac{1}{2}$ days it is back again to a position close to the sun, it follows that its angular distance from the sun increases $360° \div 29\frac{1}{2} =$ about $12°$ per day. Now, as it takes the earth 50 min. to turn through this angle, it is not surprising to find the moon crossing the meridian about 51 min. later each day. (The extra minute is due to the fact that, during the 50 min., the moon moves a trifle eastward.) Just as we define a solar day, the interval between successive passages of the moon across the meridian gives us what might be called a LUNAR DAY, which is $24^h50^m47^s$ of mean solar time. The reader might desire to be able to tell time by a moondial, as well as a sundial; but we can only discourage him in this attempt, for, because of the non-uniformity of the motion of the moon and the inclination of its orbit, lunar days vary in length from 24^h38^m to 24^h66^m.

Still more puzzling to moon-gazers is the variation in the time of its rising, and many of us have erred in estimating this time. We watch the full moon rise over the lake at 6:00 P.M. one night and are so thrilled by its serene beauty that we invite a friend to share the spectacle the next night. Knowing that the moon crosses the

THE MOON

meridian approximately 50 min. later each night, we arrive at lake at 6:30 P.M., to be ahead of time, and have 20 min. leeway, only to find that we have arrived too late or an hour too early—too late being the case for the autumn full moon, too early for the spring. The explanation is that, in latitudes near 45° N., the full moon following the autumnal equinox (the harvest moon) rises on succeeding nights at about the same time. In our table of the moon's declinations it is found that, at this time of the year, the last quarter is at its highest, or maximum distance north of the equator, and first quarter is at its lowest, or farthest distance south of the equator; that is, in passing from first quarter to full moon to last quarter, the moon is crossing the equator from south to north. Let us draw an analogy with the sun: When the sun is crossing the equator from south to north around March 21, it rises farther north each day, and the days become longer in the middle northern latitudes. So with the harvest moon: From night to night its rising point moves northward, and the duration of moonlight steadily increases from first to last quarter. We know that, to lengthen our day, we must either rise earlier or retire later; the harvest moon does both. It must rise so much earlier in Lat. 45° N. that it appears at about the same time; and, as a matter of fact, in still more northerly latitudes, it rises earlier each evening. We leave the reader to investigate the spring full moon.

THE direction of the horns of the moon is an excellent subject for argument. Suppose that in Lat. 40° N. the sun is setting in the west, as it always does, and the date is either March 21 or September 22. The sun is then on the equator and traces for us the celestial equator of the sky; hence the sun cuts the horizon at an angle of 50°, for this is the angle of the celestial equator to the horizon at Lat. 40° N. New moon for either date is near the equator; but, again referring to our table of lunar declinations, around March 21 the first quarter is far north of the equator and, on September 22, far south. Therefore, when the moon is a few days old, it is north of the celestial equator on March 21, south on September 22. In fact, the line connecting the setting sun and the new moon on or about the time of the vernal equinox is inclined in Lat. 40° N. only 11°–21° to the vertical. Consequently, upon using the rules previously mentioned for horns, we discover that these lunar appendages for the spring moon are directed upward. We then say that the horns "hold water," and hence the appellation "wet moon." On the other hand, as the reader can readily verify, the autumn crescent moon is "dry" and sets almost immediately after the sun.

WHILE the moon is traveling around the earth and changing its phases, the earth and moon are traveling about the sun. Perhaps some of us have already concluded that the path of the moon about the sun is a wavy line—an erroneous conclusion, for the moon's orbit is always concave to the sun. If we drew the paths of

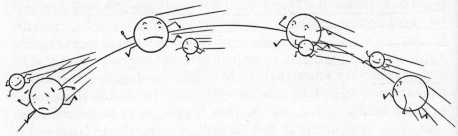

earth and moon to scale on this page, we could not distinguish between the two, since the moon's distance from earth is only one four-hundredth the sun's. Let us, instead, visualize their motion about the sun. What a celestial race the earth and moon run, as they whirl about the sun at an average speed of $18\frac{1}{2}$ mi. per sec.! At full moon we might say they are running neck and neck with the moon on the outside track, but the moon pulls ahead with a speed exceeding that of the earth by $\frac{2}{3}$ mi. per sec. Now the moon crosses in front of the earth; and at last quarter the two—moon and earth—are racing at the same speed, with the moon a quarter of a million miles ahead of the earth. Picture what would happen if the moon came to a halt at this point, for in less than 4 hr. we would crash into it. Fortunately, the moon keeps on going; it takes the inside track and slows down, letting the earth pass it at new moon, until at first quarter the earth has the lead. Then the moon crosses over to the outside track and, gathering speed, passes the earth again at full moon. This performance is repeated twelve and a half times every year.

WE HAVE already stated that the moon's distance from the earth is about a quarter of a million miles, and skeptics have wanted to know whether the man in the moon helped us to stretch the tapeline. When we measure the distance to the moon, we use precisely the same principles a surveyor employs for geographic measurements. Those who have studied trigonometry know the procedure; but, for the benefit of those who omitted that branch of mathematics, we submit the following problem: Given a stream and two points, P and A, on either shore; determine the distance from A to P without leaving the shore at A. First of all, we select some other point, B, on the same shore as A and measure the distance from A to B. From A we sight the point P and measure the angle x that the line of sight AP makes with the "base line" AB; likewise at B we measure the angle y that BP makes with AB. Now we may draw the triangle ABP to scale. Suppose the distance AB on shore is 100 ft.; we could represent this length by a line segment AB 10 in. long on a suitable sheet of paper. By laying off angles x and y, the triangle ABP is completed; and, if the AP of drawing is 11 in., then the required distance AP is 110 ft. Trigonometry dispenses with the "to-scale" drawing; and we can compute AP, given the two angles x and y and the base line AB. In like manner, to determine the distance of the moon, we measure its angular position from two observatories of known distance apart. We find that at its closest point, or PERIGEE, it is 225,000 mi. from the center of the earth and at maximum distance, or APOGEE, 250,000 mi.

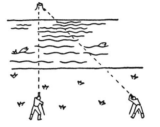 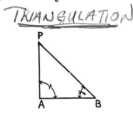

Knowing the distance and the angle subtended by the moon, it is a simple matter to determine its size. In the case of astronomical objects, the angles subtended are small; and we can develop a very simple formula connecting the size, distance, and angle subtended. The angle subtended by an object decreases as the distance increases; and for small angles, if the distance is doubled, then the angle is halved; again, if the size is doubled, the angle subtended will be multiplied approximately by 2. Therefore, the angle subtended is proportional to the ratio of the size of the object divided by its distance, being given in the formula:

$$\text{Angular diameter in minutes of arc} = 3{,}440 \times \frac{\text{size}}{\text{distance}},$$

or, upon solving for size,

$$\text{Size} = \frac{1}{3{,}440} \times \text{angular diameter in minutes of arc} \times \text{distance}.$$

Undoubtedly, the reader is wondering where we obtained the number 3,440. According to this formula, if a sphere is a mile away and subtends an angle of 1 min. of arc, its diameter is $1/(3{,}440)$ mi. Let us verify this. About the observer draw a circle a mile in radius through the sphere mentioned. Since there are 360×60, or 21,600, min. in a circle and the circumference of the circle is 2×3.1416 mi., each minute cuts off $6.2832 \div 21{,}600 = 1/(3{,}440)$ mi. on the circle, the same value as the formula gives for the diameter of the sphere. Of course, the diameter of the sphere is a straight-line segment and hence cannot be the same as the curved arc cut off. In other words, our formula is only approximate, but for angles less than a degree the error is less than a thousandth of 1 per cent. At its mean distance (237,500 mi.) the moon subtends an angle of 31 min. of arc; hence, by our formula its diameter is 2,160 mi. The diameter of the moon is thus more than one-fourth that of the earth, while its volume is one-fiftieth.

WHEN we reflect upon the size and enormous distance of the moon, we have a feeling of awe and admiration for the famous cow that jumped over it. Childhood pictures come back to us and we see a powerful cow climbing high Olympus, for this sagacious creature knows full well that friction of her non-streamlined body with

THE MOON

the denser surface atmosphere would transform her into very well-done roast beef. At the very peak she pauses to adjust her oxygen helmet and to recheck her calculations, for, upon leaving the surface of the earth, she will be at the mercy not only of the pull of the earth but also of the attraction of the moon and sun. In what direction and with what velocity should she project herself? While she soars, the moon will be moving. Then again, the earth upon  which she is standing is also in motion. Our cow is justly famous, for she has solved a complicated mathematical problem in celestial mechanics.

And now, with the moon approaching the western horizon, she takes off and flies upward with a velocity of 7 mi. per sec., the minimum velocity at which any object can escape the gravitational pull of the earth. After about 5 days, the little dog who laughed watches the cow approach her destination by means of a telescope far more powerful than any we have at present. Did she end her days by colliding with the moon, or did she whiz around it with a velocity of $1\frac{1}{2}$ mi. per sec. and then have a meteoric return to earth? The little dog's report on his observations of this interspatial flight was not sufficiently complete. From the laws of probability, however, we conclude that she missed both earth and moon and is now revolving about the sun, for we cannot credit her with strength enough to escape the gravity of our sun.

MANY of us have encountered individuals who are rather sensitive about their avoirdupois, and we have had to make use of some underhand method of discovering their weight. One way is to invite the unsuspecting victim to a gentle game of seesaw; if your partner in the game is half as far as you are from the balancing-point of the plank upon which you sit, then your partner's mass is twice yours. In general, neglecting the mass of the plank, your mass

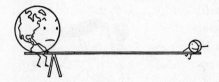

times your distance from the balancing-point *equals* the corresponding product for your partner.

The earth and the moon in their motion about the sun are playing a game of seesaw, for it is not the center of the earth but the center of mass of the earth and moon that moves in the ecliptic, obeying the Law of Areas. The CENTER OF MASS is the balancing-point, for, by definition, it is the point at which the products of mass and distance are equal for earth and moon. By "distance" we mean here the distance to the centers of the two bodies, and the center of mass is on the line through the centers. When the moon is north of the ecliptic, the earth's center is south; when the moon is south, the earth's center is north.

This bobbing up and down of the earth's center ought to be, and is, reflected in our observations of the sun and the planets. By calculations made from observations of nearby planets, we find that the center of mass of the earth and moon is about 3,000 mi. from the earth's center, i.e., approximately 1,000 mi. below the surface of the earth. The ratio of the distances from the centers of earth and moon to the center of mass is 1:80; therefore the mass of the moon is one-eightieth the mass of the earth. Since the volume of the moon is one-fiftieth the volume of the earth, it follows that, on the whole, the moon is less dense than the earth. From the principles of celestial mechanics, the subject of the next chapter, it is possible to derive other interesting data, given the size and mass of the moon. For example, the surface gravity on the moon is one-sixth that of the earth—we could lift six times the mass and jump six times as high on the moon. Modern lunar apartments could dispense with steps, for a fall of 12 ft. would be no more hazardous than falling 2 ft. on the earth.

W E NEXT look at the moon, and to bring ourselves closer to it we employ a telescope. With a good 6-in. telescope the moon appears as though it were only a thousand miles away, and we are able to see objects that are only a mile in diameter. A higher-power

THE MOON

telescope plus excellent atmospheric conditions brings us within about 200 mi. from its surface. As we watch the moon, we discover that we are always looking at that same face. Only one hemisphere can we map, for we never see the other. As the moon travels about the earth, it "turns" so as always to present the same side toward us. Though the sun and all the other planets are able to see all parts of the moon, its nearest neighbor and companion in revolution about the sun is denied that privilege. We do get a little glimpse over the poles of the moon, for its axis of rotation is inclined a trifle to its orbit. Furthermore, as the moon speeds up at perigee, obeying the Law of Areas in its motion about the earth, we see a little beyond the western limb, owing to the fact that it turns its face at a constant rate. Nevertheless, there remains 41 per cent of the moon we cannot see or expect to see for many thousands of years to come.

When the moon is a few days old, we are watching the sun rise on the western limb of the moon. Long, dark shadows are cast by its rugged surface, and from the lengths of these shadows we obtain the heights of the mountains and depths of the craters. We notice how intensely black are the shadows cast, and conclude that the moon has no atmosphere, for an atmosphere would scatter light and produce a twilight that would dimly illuminate the region near the bright crescent, just as the earth is illuminated before sunrise. Other evidence of the non-presence of an atmosphere is the abruptness with which the moon cuts off the light of a star as it passes in front of it, and the absence of a luminous atmospheric ring as the moon passes between the earth and the sun.

WHAT has become of the moon's atmosphere? If it ever did have any in the past, the evidence is that it escaped. In order to leave the earth, our famous cow found it necessary to start off with a speed of at least 7 mi. per sec. For the moon, this so-called VELOCITY OF ESCAPE is only $1\frac{1}{2}$ mi. per sec. Although this velocity is considerably greater than the average velocity of molecules of air (nitrogen and oxygen about $\frac{1}{3}$ mi. per sec.; water vapor, $\frac{1}{2}$ mi. per sec.; hydrogen, $1\frac{1}{2}$ mi. per sec.), there will always be some molecules traveling with a velocity greater than $1\frac{1}{2}$ mi. per sec.

Hence the present absence of an atmosphere for the moon does not necessarily exclude the possibility that in the past the moon had an atmosphere, which was gradually lost by molecules in the atmosphere's upper regions exceeding the speed limit of $1\frac{1}{2}$ mi. per sec. Furthermore, all known objects with velocities of escape about that of the moon or less are similarly without atmospheres, while those with velocities of escape equal to, or greater than, that of the earth have extensive atmosphere. The planet Mars, with a velocity of escape of only 3.1 mi. per sec. has an atmosphere, though a very thin one. Of course, there must be some escape of atmosphere from our own earth, but this is compensated by gas supplied by volcanic activity and by meteors. No eruptions of lunar volcanoes have been observed since the invention of the telescope; and, though the moon must receive some gas from the meteors that fall on it, evidently this source has not been sufficient to offset the loss of atmosphere by escape.

Chemists inform us that, if there is no atmosphere, there will be no surface water; and they proceed to prove their point by boiling water in a vacuum at room temperature, that is, vaporizing it without application of heat. We look at a map of the moon and read: Mare Serenitatis ("Sea of Serenity"), Mare Nectaris ("Sea of Nectar") and Mare Tranquilitatis ("Sea of Tranquillity"). But what's in a name? The chemists' surmise seems valid, for there is no evidence of water on the moon. Certain darker regions, however, appear in low-power telescopes as rather smooth plains; and it is not to be wondered at that Galileo and his contemporaries classified them as *maria*, or seas. Although the *maria* are relatively smoother than the rest of the moon, larger telescopes reveal a multitude of markings. It is because of this division of the moon into darker *maria* and lighter regions that lunar romancers are able to see the man, the lady, the donkey, or what have you on the moon.

THE best time to study the craters and mountains is when the moon is about a week old, for then the sun is rising on the central portion, and many hundreds of craters are thrown in bold relief by the shadows they cast. How different is the topography of the

PLATE 1 LUNAR "SEAS"; MARE SERENITATIS AND MARE TRANQUILITATIS

moon from that of our own earth! Whereas on the earth the mountain ridges are long and wide and made up of many approximately parallel ranges, those on the moon are short and mere thin lines of mountains, with one notable exception, the Apennines. This range extends about 400 mi. and contains some three thousand peaks, the loftiest of which rise 3 mi. above the surface. The highest lunar mountains compare in height with Mount Everest (29,000 ft.) and adjacent peaks in the Himalayas; but the lunar mountaineer's climb would not be lessened by foothills of increasing altitude, for some of the isolated lunar eminences rise precipitously from their bases to heights of 4 or 5 mi.

Far more astounding are the craters of the moon, more than thirty thousand of which have been mapped. We earthlings may be rather proud of our craters, but our nearest neighbor has craters hundreds of times the area of our largest. For instance, Kilauea, Hawaii, with its circumference of 9 mi., is a mere speck as compared with the lunar crater Ptolemy, circumference about 360 mi. Kilauea, however, is active; while all the lunar craters are dead, that is to say, they never have been seen to erupt. Imagine about a fourth of the state of Illinois walled off by a circular ridge 75 mi. in radius and about 2 mi. in height—a plain surrounded by a high wall. So appear the largest craters on the moon. Although the walls of such an imaginary crater in Illinois could be seen on a very clear day from the crater's center, on the moon the top of the crater would be below the horizon of an observer situated at the center, because of the greater curvature of the moon's surface. Smaller craters 5–20 mi. in diameter are numbered in the hundreds, and with our powerful telescopes we can see thousands of craterlets only a fraction of a mile in diameter. Many craters are so extremely narrow and deep that the sun never penetrates their bottommost depths. The floors of the craters may be below, level with, or even above the surrounding region; in fact, some are nearly filled to the brim, while others are mere holes in the plain. Numerous craters have mountain peaks rising from their interiors. Such is Theophilus, with a diameter of 64 mi., walls almost 4 mi. high, and central peaks about a mile higher than the interior. Then again, others of the larger

PLATE 2 LUNAR CRATER THEOPHILUS

craters, such as Clavius, have smaller craters on their rims and in their interiors.

Obviously, there are two ways we can account for those craters and peaks—either internal or external causes. Consequently we find one group of astronomers claiming that they are the result of past terrific volcanic action long since extinct, and that the moon is a huge cinder. Another group insists that large meteorites or swarms of meteorites weighing millions of tons have collided with the moon. The first group point to our puny volcanoes as proof of their theory; the second group describe a crater in Arizona, 1 mi. across and 600 ft. deep, which has been shown conclusively to be the work of a huge meteorite or swarm of meteorites. Then why is the moon so different from the earth? Wind and water eventually wear down the terrestrial peaks, but no such action tends to heal the scars left on the moon by past upheavals or blows. Temperature changes due to the sun and the daily shower of many millions of tiny meteors slowly pulverize the lunar rock; so we should expect to find a layer of dust lying undisturbed by winds. This type of weathering, however, takes place at an extremely slow pace.

PLATE 3 METEOR CRATER FROM THE AIR NEAR WINSLOW, ARIZONA

THE MOON

IN ADDITION to the craters and mountains, we find more than a thousand narrow, crooked crevasses of unknown depth about a mile or so in width and anywhere from 10 to 300 mi. in length. Faulting and slipping of rocks has probably occurred, and we see perpendicular cliffs on level plains, either casting a shadow on the plain or brilliantly illuminated by the sun.

As the sun rises higher in the lunar skies above the center of the moon, the shadows become shorter in this region and disappear when the moon is full. The novice looking at the moon through a low-power telescope frequently exclaims, "There's the north pole and the lunar meridians passing through the poles." The "north pole" is Tycho, a crater in the southern latitudes (a telescope inverts the image); and the "meridians" are Tycho's mysterious ray system. Many prominent craters—Tycho (diameter, 54 mi.), Copernicus (56 mi.), and Kepler (22 mi.), to name three—have long, narrow, light-colored streaks radiating from them, like the spokes of a wheel. These *rays*, as they are called, are best seen at full moon, for they cast no shadows and are level with the surface. Their maximum width is about 13 mi.; and those of Tycho extend for more than a thousand miles, approximately along the arc of a circle, uninterrupted by craters, mountains, or maria to distract the astronomer.

AS WE gaze at the moon, we wonder how things appear to the man who is supposed to live up there. We imagine ourselves as sitting on the rim of Ptolemy and looking around on the multitude of mountain peaks. Are those tiny craters ash trays? If so, the man in the moon must be an inveterate smoker. But smoking is prohibited, for there is no air. If the "ash trays" are filled, perhaps they have been filled by tiny meteors and by rocks that have crumbled off the walls. How bright the stars

PLATE 4 MOON AT AGE 16 DAYS

PLATE 5 LUNAR CRATER COPERNICUS

appear, but we can no longer call them "twinkling" stars! Truly this would be an astronomer's paradise, with no atmosphere to limit the magnification of his telescope or to diminish the brightness of the stars. Even during lunar daytime the stars are visible, for there is nothing to diffuse the blue light of the sun.

Overhead is the earth in a black sky; and there it stays, as the stars appear to travel very slowly westward behind it. It is a full earth; and we know it is midnight, for the phases of the earth give us a celestial clock for measuring the solar time of the moon. The earth is very bright and very large. It occupies almost thirteen times the area in the sky that the moon did when we viewed it from earth, and the earth's clouds and seas are far better reflectors than the lunar surface. We find no difficulty in using "earthshine" to read our lunar daily—or should we say "monthly"? (a day on the moon equals $29\frac{1}{2}$ earth days)—as we wait a week for the sun to rise. There is no news in our lunar paper, except perhaps a description of cloud movements on the earth. We turn to the weather forecast and know we can depend upon it . . . . "Fair. Extreme cold during night. Swiftly rising temperatures at sunrise, dropping rapidly at sunset. Minimum temperature $-200°$ F.; maximum at noon about $250°$ F." . . . . the forecast that we had yester-lunar-day and the forecast that we expect to have tomorrow, provided the sun is not eclipsed by the earth.

The earth, we note, is almost at last quarter; and hence the moon is near first quarter. Consequently, here on the moon directly below the earth, the sun is about to rise. There will be no dawn; but, hours before the sun appears, we will watch it gleam on the higher mountain peaks and crater walls around us. Slowly the sun rises in the east in a black sky, moving its own diameter every hour. We shield our eyes from the dazzling glare and see that not only are the earth and stars visible but there is a halo of pearly light surrounding the sun, its *corona*.

On earth, we are not so fortunate. Astronomers travel thousands of miles to view the corona, which is visible only at the time of a TOTAL ECLIPSE of the sun, when the moon passes in front of the sun and blots it out completely. If the moon moved in the eclip-

THE MOON

tic, it would cross in front of the sun every month; but the moon's orbit is inclined 5° to the ecliptic, and consequently the new moon is usually north or south of the sun. If, however, the sun is on the moon's node, the place where the moon's path crosses the ecliptic, then we shall have a total eclipse of the sun; if near the node, at the time of a new moon, the eclipse will be only PARTIAL. In fact, if the sun is even as far as 16°.5 away from the moon's node, part of the sun will be cut off by the new moon. Since it takes the sun more than a month to travel 33°, and in a month we must have one new moon and may have two, there will always be one or two partial or total eclipses of the sun for each time the sun crosses the node. But there are two nodes—one where the moon crosses from south to north and the other from north to south. If these nodes were stationary, the sun would cross them at half-year intervals; and therefore it is not unusual to find eclipses separated by about 6 mos. The nodes, however, move westward, and hence the interval is diminished to about 177 days.

Every year, therefore, we must have two solar eclipses; and we may have as many as five. As might be expected, the majority will be partial, because, for a total eclipse, the sun's center must be almost on the node at the time of new moon. Furthermore, it is essential that the moon be less than 240,000 mi. distant from earth; otherwise the angular size of the moon will be less than that of the sun, and, even though the moon's center passed through that of the sun, the outer rim of the sun would be visible. Such eclipses are called ANNULAR. For example, though the maximum number of solar eclipses (namely, five) was attained in 1935, they were all of little astronomical interest, for in no case was the solar disk completely obscured. These five eclipses for 1935 occurred as follows:

January 5: Partial eclipse of the sun, scarcely more than a mere contact and invisible in the United States
February 3: Partial eclipse of the sun, more or less visible throughout North America
June 30: Partial eclipse of the sun, invisible in the United States
July 30: Partial eclipse of the sun, invisible in the United States
December 25: Annular eclipse of the sun, invisible in the United States

Since the sun is tremendously larger than the earth, the shadows of the moon and of the earth taper to a point. At the distance of the earth, the moon's shadow is always less than 168 mi. in diameter. As the moon moves across the sun to produce a total eclipse, the moon's shadow sweeps over the earth; and, because of the moon's motion and the earth's rotation, it travels 1,000 mi. or more per hr. Anyone fortunate enough to be in the path of the shadow will see a total eclipse, weather permitting. The period of totality is at most $7^m 58^s$; but astronomers travel great distances for an eclipse lasting a couple of minutes or less, often to be disappointed by cloudy weather which makes their trip in vain. On the average, the moon's shadow falls on a particular spot of the earth once every 600–700 yr.

THE phenomenon of a total eclipse begins with first contact, when the moon appears to touch the western limb of the sun. As the minutes pass, more and more of the sun is cut off by the moon; and within three-quarters of an hour the sun appears as a thin crescent. The light of the sun now rapidly fades and takes on a weird color, owing to the difference in quality between the light near the rim of the sun and the stronger light from the interior, which now has been cut off. In the shadows of leaves we see many images of the crescent sun. Several minutes before the beginning of totality, ghostly shadow bands are seen to fall on white surfaces; and now, turning to the west, we see the moon's shadow rapidly approaching us. The thin crescent breaks up into "Bailey's Beads," which are due to the sun shining through the irregular surface of the moon.

THE MOON

But the beads last only an instant, and then the stars and the sun's corona burst into view. Never appearing twice the same, the irregular streamers of the corona extend from the sun about half the solar diameter. Frequently, flamelike eruptions, called *prominences*, rosy in color, are seen just as the visible portion of the sun disappears or just before it reappears. How rapidly those precious minutes of totality seem to pass! Yet, during this short interval, the observational astronomer must make all his observations of the corona or whatever other information he seeks, such as: the search for new planets near the sun; determination of the exact relative position of the moon and sun; and the shift, if any, in the apparent position of stars near the sun, predicted by Einstein in his Theory of Relativity. Consequently, each second of the period of totality must be accounted for; and, weeks before the eclipse occurs, the astronomers make practice observations, so that there will be no time-consuming mistakes. Only too soon Bailey's Beads reappear and the total eclipse is over.

WHEN the moon passes into the earth's shadow, we have an eclipse of the moon. If the sun is on one node of the moon's orbit, then the earth's shadow will fall on the opposite node; hence the conditions for an eclipse of the moon are that the moon be full and, as was the case for solar eclipses, the sun near a node. Lunar eclipses are not so frequent as solar: we may have no eclipse whatsoever in a year, and we may have three at most. Furthermore, the earth's shadow does not exclude all light, for earth's atmosphere "bends" some of the sunlight to fall on the moon. The dull, reddish appearance of the moon at totality is due to light that has passed through our air.

The problem of determining accurately the time, duration, and magnitude of a solar or lunar eclipse is not simple. Nevertheless, the predicting of solar and lunar eclipses was one of the chief duties of astronomers in China two thousand years before Christ. In fact, the Chinese claim the first record of a total solar eclipse; and their emperors traced their ancestry to the sun, styling themselves "The Son of the Sun." To insure accuracy, ancient Chinese astronomers were relieved of their heads if they erred in their calculations. Though the more refined and reliable methods of celestial mechanics have replaced the procedure of the ancients, their scheme is not without interest.

By studying records of past eclipses, the early astronomers of China and Chaldea discovered that eclipses did not occur haphazardly, but usually at regular intervals of 18 yr. plus 10 or 11 days— THE SAROS, as it was named, meaning "repetition." We trace this same cycle in recent total eclipses: January 14, 1907; January 24, 1925; February 4, 1943 (predicted). During the Saros there occur about twenty-nine lunar eclipses, thirty-one partial eclipses of sun, and ten total eclipses of sun. Each of these eclipses is usually repeated 18 yr. and 10 or 11 days later but is then visible in longitudes 120° farther west. That this method is not infallible we infer from the sudden demises of two Chinese astronomers, Ho and Hi, immediately following the solar eclipse of 2169(?) B.C.

We, for our part, courageously list the total eclipses which will be visible in the United States for the next 50 yr.:

1945 July 9	Northwestern states
1954 June 30	Northwestern states
1963 July 20	Northeastern states
1970 March 7	Florida
1979 February 26	Northwestern states

and we refer the reader to Oppolzer's *Canon der Finsternisse* if he desires information about any eclipse that has occurred or will occur in the interval 1207 B.C. to 2162 A.D. This voluminous work is of great use in fixing historical dates from records of eclipses.

WE CONSIDER next the effect of the moon on the earth. Some heat and light are received from the moon, about as much in a year as from the sun in 13 sec. Then, again, our weight is a trifle less when the moon is overhead than when it is on the horizon, for, when it is overhead, it is pulling on us with a greater force per unit mass than on the center of the earth; in other words, there is a slight tendency to pull us away from the earth's center. The

THE MOON

same effect will take place when the moon is directly below us, for then it tends to pull the earth's center from under our feet, again making us lighter; the variation in weight, however, is only one part in several million. Nevertheless, this slight difference in acceleration at center and at surface is the major cause of our ocean tides.

If the earth were not rotating, the oceanic waters directly below the moon would be piled up a couple of feet, and practically the same upheaval would take place on the opposite side of the earth. As the earth rotates, these high points remained fixed relative to the moon (though carried forward a little by friction and other causes) and finally are met by the shore. Consequently, we find there are two tides daily, on the average a tide every 12 hr. and 25 min., which is one-half the interval between successive passages of the moon across the meridian. The sun likewise produces tides of lesser magnitude, which may oppose, or add to, the lunar. The actual magnitude of a tide depends not only on the sun and moon but on the depth of ocean, shape of shore line, barometric pressure, winds, etc. Nevertheless, by comparing the observed tides with the positions of the sun and moon, relations are obtained by which the magnitude and time of future tides can be predicted.

These tidal actions act as a brake on the rotation of the earth; hence we might expect that eventually the earth will imitate the moon in the matter of always presenting the same face toward it. But we have forgotten the sun's tidal action. Which will win out? Our disregard of all other effects makes any deduction a mere speculation, but certain experts conclude that it will be the sun and that the moon will come toward the earth, to be finally torn apart by the earth's attraction—thus terminating the study of the moon.

CHAPTER 5

CELESTIAL MECHANICS

> When Newton saw an apple fall, he found
>   In that slight startle from his contemplation—
> 'Tis *said* (for I'll not answer above ground
>   For any sage's creed or calculation)—
> A mode of proving that the earth turn'd round
>   In a most natural whirl, called "gravitation";
> And this is the sole mortal who could grapple,
> Since Adam, with a fall, or with an apple.
>
> Man fell with apples, and with apples rose,
>   If this be true; for we must deem the mode
> In which Sir Isaac Newton could disclose
>   Through the then unpaved stars the turnpike road,
> A thing to counterbalance human woes:
>   For ever since immortal man hath glow'd
> With all kinds of mechanics, and full soon
> Steam-engines will conduct him to the moon.
>   —*Don Juan*, Canto x, stanzas 1–2

IT IS somewhat surprising that "the fruit of that forbidden tree whose mortal taste brought death into the world" should obtrude itself into legends concerning so abstruse a subject as celestial mechanics. We have all heard the story. An apple suddenly became conscious of a gravitational urge and descended with increasing velocity upon a sleeping mind, awakening the brain cells to an understanding of the glory of the universe. For our part, how-

CELESTIAL MECHANICS

ever, we question very much the authenticity of this tale and doubt that there ever existed a man who could, after such a severe awakening, calmly contemplate the mathematical and physical implications of the planetary and lunar trajectories. This *canard* ought to be thrown out of court, for that stupendous induction, the Law of Gravitation, was not a hit-or-miss proposition—the real account is one of the most perfect examples of the methods of scientific procedure.

ALL stories have a beginning, and ours begins thousands of years before the birth of Christ. Nevertheless, even at the present moment, we have not nearly reached its conclusion. Science—and now we mean *true* science—starts with observations. Picture, then, an ancient Chaldean shepherd of Asia or a pyramid-builder of Egypt looking up at the skies and pondering the why and the wherefor. He discovers that there is some sort of order in the skies: He sees objects rise in the east and travel across the heavens; they repeat this same performance the next day and the next and the next, and so on throughout his lifetime. He relates his observations to his sons, who in turn gaze at the sky and discover the same, as well as other hitherto unnoticed, phenomena. They watch the sun, that brilliant and awe-inspiring disk, describe its apparent yearly circle among those scintillating points of light, the stars. They follow that object we call the "moon" in its rapid flight and changing phases. They notice that practically all the points of light move together across the heavens, keeping the same arrangement; but their attention is drawn particularly to a few points which seem to wander among the stars.

These points they called "wanderers," for the word PLANET means a wanderer. Just as the moon and sun selected an eastward direction for their motion, so the planets seemed to prefer to roam eastward; but, while the sun and moon appeared to travel uniformly, the planets were somewhat erratic in their movements. Mercury and Venus wandered from one side of the sun to the other, always keeping rather close to the sun; Mars would be traveling eastward at about half the speed of the sun, passing through the zodiacal constellations, when, for no apparent reason, it would slow down, stop, go back for some distance, and then resume its eastward motion. The same phenomena were remarked in the two remaining planets that they could see—Jupiter and Saturn.

Throughout the ages these five planets have pursued their peculiar paths; and, once the reader is able to identify them, he may observe for himself their long eastward or *direct motion* and the comparatively shorter westward or *retrograde motion*. The wandering of Jupiter is manifest in its EPHEMERIS (a table showing calculated positions at regular intervals) for 1934.

EPHEMERIS FOR JUPITER

Date	Declination	Right Ascension	Date	Declination	Right Ascension
Jan. 1........	−7° 3′	13^h20^m direct	July 1........	− 3°58′	12^h53^m direct
11........	7 22	13 24 direct	11........	3 57	12 56 direct
21........	7 35	13 26 direct	21........	4 3	12 59 direct
Feb. 1........	7 21	13 28 direct	Aug. 1........	5 33	13 4 direct
		stationary	11........	6 8	13 9 direct
11........	7 40	13 28 retrograde	21........	6 46	13 15 direct
21........	7 32	13 27 retrograde	Sept. 1........	7 31	13 23 direct
Mar. 1........	7 20	13 25 retrograde	11........	8 14	13 30 direct
11........	7 1	13 22 retrograde	21........	8 59	13 37 direct
21........	6 36	13 19 retrograde	Oct. 1........	9 44	13 45 direct
Apr. 1........	6 6	13 14 retrograde	11........	10 30	13 53 direct
11........	5 37	13 9 retrograde	21........	11 16	14 1 direct
21........	5 8	13 4 retrograde	Nov. 1........	12 5	14 10 direct
May 1........	4 43	13 0 retrograde	11........	12 48	14 19 direct
11........	4 22	12 56 retrograde	21........	13 30	14 27 direct
21........	4 7	12 53 retrograde	Dec. 1........	14 9	14 35 direct
June 1........	3 58	12 52 retrograde	11........	14 46	14 43 direct
11........	3 57	12 51 stationary	21........	−15 19	14 51 direct
21........	−4 3	12 52 direct			

The direction of Jupiter's motion among the stars is obtained from the column "Right Ascension." Ephemerides such as this, but far more complete and accurate, are given for past, present, and future years in *The American Ephemeris and Nautical Almanac;* and a study of these tables reveals that, in a planet's retrograde motion, it may at times cross its own path to make a loop, while on other occasions we observe a zigzag motion, the planet not crossing its own path. The position of the retrograde motion among the stars changes continually, but in every case the planets are always close to the ecliptic—a fortunate situation for the neophyte.

CELESTIAL MECHANICS

THE question in ancient times was how to account for this motion. Though the Eastern peoples were indefatigable observers, it was left for their Western neighbors to attempt to answer these questions. The stars, the sun, and the moon seemed so orderly that the Greeks and their mentors, the Egyptians, suspected the behavior of the planets was not really erratic but the result of combinations of orderly motion. In venturing a logical description of the motion of the heavenly bodies, the majority of the Greek philosophers assumed the earth to be stationary. A few, as we have already noted, claimed that both earth and planets revolved about a stationary sun; Pythagoras (569–470 B.C.) and Aristarchus (310–250 B.C.) were of this school. The absence of an observable parallax, however, and the consideration that a moving earth seemed so contrary to human experience, caused such great thinkers as Hipparchus (190–120 B.C.) and Claudius Ptolemy (A.D. 100–170)—who argued that, if the earth were moving, falcons and other birds would be left far behind—to start with the assumption that the earth was fixed and at the center of the universe. The appeal of such an assumption is great, for, with a little more assuming, the individual becomes the center of everything, with the entire universe revolving about him! Spheres and circles were such perfect surfaces and curves that, in the geocentric (earth-at-center) plans proposed, the planets were situated on transparent crystalline spheres, and all motions were combinations of motions in a circle.

A SCHEME, giving all the irregularities that the ancients could discover in the planetary motions, was finally developed; and so ingenious was their method that the designers of so-called "amusement rides" might well profit by a study of the Ptolemaic system. Imagine a Ptolemaic-Martian-Merry-go-round, a composite of two carousels, one about two-thirds the diameter of the other, the center of the smaller (or, in the terminology of the time, the *epicycle*) being placed at the edge of the larger (or the *deferent*). We poise Mars on the rim of the epicycle and start both merry-go-rounds revolving in the same direction, with the angular velocity of the epicycle twice that of the deferent. If this roundabout were so constructed that we could stand at the center of the deferent without participating in its motion, then, as we watched Mars whirl around, we would comprehend the Ptolemaic system. When Mars was closest to us, the faster rotation of the smaller circle would cause the planet to retrograde; but, at greatest distance from us, the direct motion of the planet would attain its maximum.

The complete Ptolemaic Merry-go-round would have pairs of roundabouts for each planet similar to the one described for Mars, each deferent turning about the earth with its own angular speed; but the sun and moon would move only in single circles. Of course, an exact reproduction would be more complicated in that it would require our disks to be replaced by clear crystalline spheres, with an outer sphere holding the stars and carrying along with it, in its daily turning, all the other spheres. Then, as the spheres clanked on each other, we too would hear the "Music of the Spheres"—a privilege that was reserved only for the gods and the extremely pious. We shall not, however, enlarge upon this theory of Plato, for the music of merry-go-rounds is a painful subject.

Let us hasten to say that we are not venturing to ridicule the ideas of Eudoxus (fourth century before Christ), the inventor of the crystalline spheres, or those of Hipparchus, the Newton of antiquity. It might well be that the original theorists did not intend their speculations to be taken as the true physical description of the universe, but rather as a mathematical device for predicting the position of the planets. Ptolemy's great work, the *Almagest*, as it was called by the Arabians in the translation they made (A.D. 827), meaning "the greatest," was the textbook of astronomers for many centuries.

FOR almost fourteen hundred years the Ptolemaic system went unchallenged. Then he whom we have already mentioned, the Polish monk Copernicus (1473–1545), reverted to the ideas of Pythagoras and Aristarchus, insisting in his *De revolutionibus orbium caelestium* that these observed planetary motions might be the result of a motion of the earth as well as of the planets themselves. He showed how all observable motions predicted by the Ptolemaic

CELESTIAL MECHANICS

scheme could also be obtained by placing the sun at the center of the system and supposing the earth and the remaining planets to be revolving uniformly about the sun in circles. Some of us might have difficulty in understanding how motion in a circle always in the same direction about the sun will cause a planet such as Mars to move occasionally in the westward or retrograde direction, until we remember that this motion is being observed from the earth. We can plot corresponding positions of the earth and Mars and thus discover the way the direction of its apparent motion changes, or we might

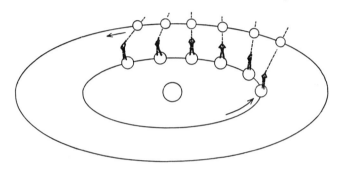

compare this phenomenon with other experiences. Now, according to the Copernican system, the closer a planet is to the sun the faster it travels. Thus the earth travels faster than Mars; and, when passing Mars, it is like riding along in a fast train and overtaking a slow train. As the limited (earth) passes the local (Mars), the local appears to go in the reverse direction for the passengers of the limited.

Copernicus' chief argument for his system was its simplicity. Perhaps the scheme was simpler geometrically, for it involved fewer circular paths. Nonetheless, degree of simplicity is a matter of viewpoint and not a safe criterion for judging theories. After all, *is* a rapidly spinning earth, whirling about the sun, a simple idea? The proper procedure is to seek observations that will differentiate between the two theories. Now, in our hypothetical Ptolemaic Merry-go-round, the epicycles for Venus and Mercury should have their centers on the line joining earth and sun, and these planets would always be on the same side of the sun as the earth. Inasmuch as they shine only by light coming to them from the sun, merely a small portion of their illuminated halves would be visible. To put it another way, these planets, if spherical in shape, would always appear as thin crescents when viewed through a telescope, provided, of course, the theory of Ptolemy was valid. But in the Copernican system it is possible for all planets to pass behind the sun relative to the earth and, in so doing, they would present their illuminated halves to us and be seen in gibbous phase. So Copernicus boldly announced that, if human sight could be sufficiently enhanced, then these two planets—Mercury and Venus—would be seen to exhibit the same phases as the moon.

IN 1609 the great Florentine physicist Galileo Galilei (1564–1643) learned of the construction of a telescope by Jan Lippershey, a Dutch spectacle-maker. Lippershey ignored the astronomical possibilities of his invention; so we may credit Galileo with the first astronomical telescope—a piece of lead pipe with a lens at either end. Though reared in the ideas of Ptolemy, he became a staunch supporter of the heliocentric (sun-at-center) system. His telescope, a cudgel for silencing the disciples of Ptolemy, was destined to revolutionize astronomy.

On January 8, 1610, Galileo turned his lead pipe toward Jupiter and saw what he regarded as a miniature Copernican system—the four larger moons of Jupiter circling the planet. What *more* proof could man ask? Here in the sky was a model of the solar system! The news spread rapidly but was not generally believed. Dissenters admitted that doubtless a telescope behaved itself when directed to terrestrial objects, but claimed that, when applied to the heavens, it immediately became an instrument of Satan. In order to justify this contention and make their arguments more convincing, a few monks even put paper devils in front of telescopic lenses. (It is curious to note that, even today, the Galilean telescope is used chiefly for viewing what, to some, are still considered abominations of the devil; we refer to theatrical performances, for the opera glass is of the same optical design as the first telescope. Modern astronomical telescopes of the refractor type give us an inverted image and therefore are not desirable for terrestrial purposes.)

All this persecution had little effect on Galileo; and, upon the decease of one of the many scientists who had declined an invitation to look through his telescope, he generously said, "I hope that he saw them on his way to heaven"—referring, of course, to the moons. When Galileo directed his attention to the planet Venus, the Ptolemaic theory received its first real blow. He saw the planet first in the gibbous phase; but, being cautious, he announced his discovery in the anagram: *Haec immatura a me iam frustra leguntur. o y* ("These unripe things are now read by me in vain"). When finally he viewed the planet in crescent form, he translated the anagram by interchanging letters into: *Cynthiae figuras aemulatur mater amorum* ("The Mother of the Loves imitates the form of Cynthia"), or, in plain English, "Venus passes through the same phases as the moon." Thus was the prediction of Copernicus fulfilled.

ALTHOUGH the observations of Galileo refuted the theory of Ptolemy, nevertheless it still is possible to account for these observations by a geocentric theory, that is, one in which the earth is stationary. Such a scheme was proposed by the Danish astronomer Tyghe Brahe—in the Latin form, Tycho Brahe—(1546–1601), prior to Galileo's observations of Venus. In order to explain the Tychonic system, we again resort to the merry-go-rounds. As with the Ptolemaic Roundabout, the sun circles the stationary earth; but, instead of having different centers for the epicycles of the planets, in Tycho's scheme all the planets, except the earth, circled the moving sun. In our Tychonic Carousel, Venus is to be placed on a roundabout with sun at center, and this mechanism is then placed on a larger merry-go-round with earth at center. Consequently, the planet would pass both behind and in front of the sun and would show all the phases of the moon. Many authorities regard Tycho's theory as a backward step, for it involved a stationary earth. Were we not prejudiced by later discoveries, we might be inclined to regard the Tychonic system as simpler than, and superior to, the Copernican, for it contains the same number of circular motions and predicts the same observations, with one important exception—parallax. According to the Tychonic scheme, no parallax should be observed; and, since neither Tycho nor any observer prior to his time had observed parallax, he was quite justified in his procedure.

But even the schemes of Copernicus and Tycho were not without their troubles. A few hundred years before the time of Copernicus, astronomers were noting differences between the theoretical Ptolemaic paths and the observed motions. They attempted to modify the Ptolemaic system by adding more epicycles; thus they would have a planet move in a circle, the center of which moved in a second circle, the center of the second moving in a third, and so on. In his theories even the mighty Copernicus likewise was a victim of this method of combining circular motion. In fact, any motion may be resolved in this fashion, if one permits a sufficient number of circular paths. But, as to the complexity of the result, we quote the remark of Alphonso the Wise (1221–84) concerning Ptolemaic astronomy: "Had I been present at the Creation, I would have given some useful hints for the better ordering of the Universe."

DESPITE his so-called "backward" step, Tycho's painstaking observations of the planets were one of the important links in the chain of events that led to the Law of Gravitation. An aristocrat by birth, Tycho was so impressed in his fourteenth year by the fulfilment of the prediction of an eclipse that he chose the then-considered plebeian occupation of an astronomer and, we grudgingly add, astrologer. His instruments at the Uranienborg Observatory, a gift from the king of Denmark, were the most accurate ever made before the introduction of the telescope.

This was the day when those intricate and beautiful astrolabes, which we admire in museums, were being replaced by the more practical and precise quadrants. In richly ornamented robes Tycho measured the altitudes of planets and stars with a quadrant, sighting along its edge—just as we aim a rifle—and determining the inclination of the instrument by means of a plumb line moving in a graduated 90° arc. He measured the maximum altitude of planets and the time of the meridional passage. These observations he translated into right ascension and declination and presented to one of his

CELESTIAL MECHANICS

students, a young German mathematical astronomer, Johannes Kepler (1571–1630). It was Kepler's herculean task to account for those departures observed by Tycho in the motions of the planets, particularly of Mars, which were not in accordance with Tychonic, Ptolemaic, or Copernican theories. The labors of Kepler were the consequence of this assignment.

Kepler had had some correspondence with Galileo, who, being a staunch supporter of the ideas of Copernicus, had probably directed Kepler's thoughts toward the heliocentric theory (a stationary sun). Convinced that Tycho's observed departures from all theories so far proposed were *real*, and believing that the observed complicated motions were a result of simpler planetary motions, including that of the earth, he proceeded to try out other curves for the paths of the planets. Now, in their study of geometry, the Greeks had followed the straight line and the circle by another simple closed curve, the ellipse; and therefore it is not surprising to find Kepler, after discarding nineteen other hypotheses, investigating the possibilities of elliptical orbits.

Where should he place the sun in these ellipses? With no other knowledge, we should probably guess, as did Kepler, that the sun should be at the center. Then arose such questions as: How should the earth and planets move in these ellipses? How incline the various planes? What should be the longer and shorter axes of the ellipse for earth, if earth were a planet? What should be their orientation? There was no direct solution for these questions. All Kepler could do was to experiment with various combinations; and for years he toiled, but to no avail.

UNDAUNTED, he tried again, this time placing the sun off center at one of the foci of the ellipse; and finally, after a total of nearly twenty years of almost incredible labor, he found that he could account for Tycho's observations by means of the following three simple laws:

I. ELLIPTICAL ORBIT LAW: Each planet moves in an ellipse, with the sun at one of its foci.

II. LAW OF AREAS: The line joining each planet to the sun sweeps over equal areas in equal intervals of time.

III. HARMONIC LAW: The cubes of the mean distances of any two planets from the sun are to each other as the squares of their periods of revolution about the sun.

The first two laws we have already encountered in our discussion of the motion of the earth about the sun. The third law perhaps requires some explanation. By MEAN DISTANCE OF A PLANET is meant one-half the sum of its maximum (aphelion) and minimum (perihelion) distances. Given two planets with periods of revolution about the sun p and P, respectively, and mean distances a and A, then

$$\left(\frac{p}{P}\right)^2 = \left(\frac{a}{A}\right)^3.$$

We can always take the earth as one of these planets and measure time in years and distances in astronomical units; therefore, let us consider P equal to 1 yr. and A equal to 1 astronomical unit in the previous relation. Since p divided by 1 is p, and a divided by 1 is a (division by unity is one of the most important of mathematical operations), it follows that Kepler's Third Law may also be expressed in the mathematical form

$$p^2 = a^3,$$

or in words—

The square of the period of a planet in years equals the cube of its mean distance in astronomical units.

Given the mean distance of any planet, this law enables us to compute its period; and conversely, given its period, the mean distance is determined. For example, observations plus calculations assigned to the recently discovered planet Pluto a mean distance of 39.6 astronomical units. Consequently, if p is the period of Pluto in years, then

$$p^2 = (39.6)^3 = 39.6 \times 39.6 \times 39.6$$
$$= 62,099.$$

CELESTIAL MECHANICS

Since the square of p is 62,099, the period of Pluto's revolution about the sun is the square root of 62,099, or 249 plus a fraction years.

Kepler himself was elated over the discovery of this last law, as evidenced by his own remarks: "The die is cast, the book is written, to be read either now or by posterity, I care not which. It can await its reader. Has not God waited six thousand years for an observer?"

We frequently criticize Kepler for his many wild speculations in astronomy, as well as in astrology; but even today, with the aid of all the mathematical machinery developed in the last three hundred years, it still would be an extremely difficult problem to rediscover Kepler's laws from Tycho's observations. True, Kepler did propose a system of celestial harmonics with Saturn and Jupiter taking the bass, Mars the tenor, Earth and Venus the counter, and Mercury the treble, Earth's part consisting of *mi, fa, mi*, which is to be interpreted as "misery, famine, misery"—but those same eccentricities, which have called forth so much adverse criticism, might have been essential in solving his monumental problem. It is comforting for the lowly mind to regard Kepler's more popular treatise on astrology as an example of how even the mighty intellects have been induced to prostitute their science in order to subsist. Despite his strayings from the path of cold reason, his three simple laws delivered the coup de grâce to the Ptolemaic system and cleared the way for celestial mechanics.

KEPLER'S theory, as well as the other theories mentioned so far, might be classified as geometrical theories; that is, they are geometrical descriptions of how the observations are produced. His three laws do not give any clew as to *why* the planets describe such paths. Perhaps we should not hope to answer fully the question "Why?" when applied to the behavior of nature, but it would be more gratifying to account for these motions in terms of everyday experience.

It required the genius of Sir Isaac Newton (1642–1727) to translate Kepler's induction into a physical law. Born the year

Galileo died, Newton took his degree at Cambridge in 1665, giving evidence of extraordinary ability in the mathematical sciences. Upon completing his scholastic training, he retired to the country to escape the plague which was raging in London and vicinity. It was there, "far from the madding crowd," that he discovered the binomial theorem of algebra; founded a new branch of mathematics, which he termed "fluxions" but which is now better known as the "calculus"; developed a theory of light; and formulated those very laws of motion which we have already encountered in discussing the rotation of the earth. Each was a notable contribution to science. His genius was quickly recognized; and, while yet a young man, Newton received a professorship at Cambridge University.

In the study of motion of bodies at the surface of the earth, we utilize the system of mechanics built up by Newton. The basis of this system is the ordinary axioms of geometry *plus* Newton's three Laws of Motion (p. 8). If we desire to know the path of a bullet or the behavior of a spinning top, we start with this basis, expressing it usually in the form of equations and arriving at the solution of our problem through mathematical operations.

NEWTON, however, did not limit himself to terrestrial phenomena. He postulated that those same laws which were valid for objects on the earth applied likewise to celestial bodies. Observations indicated that the planets obeyed Kepler's law; hence he sought the implications of Kepler's three laws from the standpoint of his own postulates. His findings form part of his immortal work *Philosophiae naturalis principia mathematica*. Written in 1685 and 1686—thanks to the insistence of his friend Edmund Halley—*The Principia*, as it is commonly known, was published in 1687. This imposing treatise is well described by the words of the brilliant German philosopher and mathematician Leibnitz (1646–1716), a rival of Newton in that he independently invented the calculus: "Taking mathematics from the beginning of the world to the time when Newton lived, what he had done was much the better half."

Unfortunately, in one of the thrilling parts of our story, we must omit a considerable number of pages—pages that are covered

CELESTIAL MECHANICS

with beautiful mathematical logic requiring a knowledge of the calculus to be fully appreciated. In order to preserve the continuity of our tale, however, we shall employ a tiny arrow to indicate the mathematical operations of solving differential equations or complicated geometrical problems and list the inferences Newton obtained from Kepler's three laws in conjunction with his own Laws of Motion.

a) Law of Areas+Laws of Motion → Planets are acted upon by a force which is always directed toward the sun.

b) Law of Areas+elliptical orbits
+Laws of Motion → This force varies inversely as square of distance from sun.

c) Harmonic Law+Laws of Motion → The force is proportional to the mass of the planet.

In passing, let us note that the existence of a force on the planets immediately follows from the First Law of Motion; were it not so, the planets would continue to travel indefinitely in a straight line with a constant speed. The proof that elliptical orbits, sun at focus, imply that this force varies inversely as the square of the distance from the sun must be reserved for treatises on celestial mechanics.

FURTHERMORE, Newton did not stop with a mere statement of the immediate consequences of Kepler's laws, but proceeded to state a law of universal gravitation, namely,

Every particle in the universe attracts every other particle with a force proportional to the product of the masses divided by the square of the distance.

Let us examine more critically this last step. In Newton's statement of the Law of Gravitation, he is inserting *more* than he deduced from Kepler's laws, which implied only that there was a gravitational force between each planet and the sun. Newton, however, claims that such a force exists between *every* particle, the Keplerian planetary motions being only *one* instance of the action of this force. Newton's law immediately challenges us to seek other instances.

RETURNING somewhat reluctantly from an enjoyable vacation in the country and lacking the assistance of a porter, we become bitterly conscious of this merciless and relentless world. We *know* that each and every particle of the earth is pulling, and will continue to pull, on each and every particle of our valise—to say nothing of the attraction of sun, moon, planets, and distant stars. We raise a hand to mop our forehead, well aware that this rash action immediately alters the course of Mars, but we are past caring. There is no comfort in knowing that the accumulated attraction of the particles of the earth on our burden would be the same, if we concentrated them all at the center of the earth four thousand miles below us; nor do we derive any solace from the thought that a super-stratopheric flight eight thousand miles above the surface of the earth would diminish the earth's pull to one-ninth its value.

But this mysterious force, gravity, is of some service, holding us to this spinning globe and preventing the earth from leaving our source of energy. A butcher weighing a chop should thank gravity for its constancy; and, as we watch his procedure, we hope he appreciates the principle underlying his operations. Just as he utilizes gravitational attraction to obtain the mass of the chop, so the physicist employs it in his determination of the mass of the earth; while the astronomer carries it beyond to measure the mass of sun, planets, and stars. We watch the butcher's actions carefully as he compares the attraction of the chop and a given mass, the earth, assuming this celestial body to be his place of business, with the attraction of a known mass and the same given mass. If he were on the moon or any other body, his method and results would be the same, though his given mass would be different.

The physicist proceeds as the butcher; but, since his "chop" is the earth itself, he must find some other given mass. Theoretically, any mass would suffice; but to simplify his calculations the physicist borrows a small lead ball. First he measures the attraction between

CELESTIAL MECHANICS

the earth and this given lead ball, then with a specially designed balance, the torsion-balance, he finds the extremely minute attraction between the given lead ball and a ball of known mass, noting at the same time their separation. So feeble is this force that it was not until 1798 that Lord Cavendish succeeded in measuring the mass of the earth with his torsion-balance. The attraction between two iron balls weighing a ton each and almost in contact is one ten-thousandth the pull of the earth on a pound of steak.

The remainder of the physicist's problem is one of computation. In this respect the butcher is more fortunate; *his* chop and known mass are situated at the same distance from the center of the given mass (the earth)—if attraction of the chop on a given mass equals that of a known mass on a given mass, then it follows from Newton's Law of Gravitation that mass of chop and known mass are equal. The physicist, however, must take into account the distance from the earth's center to a given lead ball, as well as the distance between the centers of a given ball and a ball of known mass; for gravity varies inversely as the square of the separation. We omit the arithmetic and state the physicist's result: The earth's mass is 13,200,000,000,000,000,000,000,000 lb. Despite our familiarity with large numbers, this figure is probably unintelligible. More comprehensible is the statement that the earth is five and a half times as massive as a sphere of water of the same radius.

THE astronomer has no balance-scales similar to the butcher and the physicist to weigh the sun, planets, and stars. He must resort to other ways of finding the attraction of these masses. To illustrate his method, we fill a pail with water and swing the bucket around in a vertical plane. If the angular velocity of our arm is sufficient, the experiment should have no disastrous consequences. Calculations based on the laws of motion show that, for a normal arm, the centrifugal force will exceed the pull of gravity if the bucket makes at least one revolution every 2 sec. (a lower angular speed will result in a shower; on the moon the smaller gravitational pull would permit us to swing the pail more leisurely). The minimum speed for safety depends, of course, on the arm; but, given gravity and arm, the problem is readily solved. Conversely, our pail of water enables us to measure the gravitational pull, given length of arm and angular speed that will just keep the water in the pail.

In like manner, the pull of the earth's gravity on the moon is just sufficient to keep it in its present more or less circular orbit; consequently, we can readily find the pull of the earth on the moon. By a similar attack we derive the attraction of the sun on the earth. Knowing these attractions, our method follows that of the butcher and the physicist, and we conclude that the mass of the sun is more than three hundred thousand times that of the earth. The procedure is quite general—the mass of planets with satellites and the combined masses of two stars revolving about each other are easily determined, provided we know their periods of revolution and their mean separation.

The reader should note that we have been concerned here with the *mass* of bodies—not their *weight*. Though we ordinarily derive mass from the attraction the body exerts on another body, mass is a property of the body itself independent of its position. Weight, however, is the magnitude of the attraction; for terrestrial objects, it is the pull of the earth's gravity. Thus a man of considerable avoirdupois can reduce his weight to one-sixth its value by living on the moon, but his mass is invariant unless he changes his diet.

THE gravitational method of determining the mass of a planet is not limited to planets possessing satellites. The Law of Gravitation states that every particle attracts every other particle; consequently, the particles of one planet pull on the particles of every other planet, and conversely. This mutual attraction is small in comparison with the attraction between sun and planet, for it produces only slight departures from Keplerian motion. Nevertheless, the study of these deviations from Keplerian orbits, or PERTURBATIONS, as they are called, enables us to find the mass of moonless planets. The story of the discovery of Neptune, one of the great triumphs of celestial mechanics, which is being reserved for the next chapter, illustrates how perturbations enable us to establish the existence and location of a previously unnoticed mass. In the two

CELESTIAL MECHANICS

centuries that have elapsed from Newton's time, thousands of such departures or perturbations have been predicted, and later verified by observations made with modern instruments. For example, the earth is not moving in an ellipse of constant shape and orientation; its path is being perturbed so that, for the next twenty-four thousand years, it will become more and more circular, thereby reducing the inequalities in the lengths of the seasons.

Perhaps the reader is also perturbed and wonders how Newton's induction could be derived from Kepler's laws, when universal gravitation demands that the planets deviate from Keplerian paths. Although the Law of Gravitation was "discovered" through Kepler's laws, a little more was added to its formulation, namely, that *every particle attracts every other particle*. It was this very slight addition which caused the French scientists to hesitate a century before accepting the generalization of Newton; but, when it was finally adopted in France, they became leaders in Newtonian celestial mechanics through the researches of such great mathematicians as Lagrange (1736-1813), Laplace (1749-1827), Leverrier (1811-77), and H. Poincaré (1854-1912). As a matter of fact, if perturbations had not been observed, the Law of Gravitation would have been abandoned; but, on the other hand, it is fortunate that these perturbations were not discovered by Tycho. If they had been, the subsequent chain of events—Kepler's formulation of his law and Newton's generalization—might not have been forged and the Law of Gravitation might have remained undiscovered up to the present day.

THE extension of the principles of gravitation to include stellar motions requires further justification. From the laws of motion it follows that two spherical bodies, revolving about each other and subject only to their mutual gravitation, must move in circles or ellipses about their common center of mass. This is called the PROBLEM OF TWO BODIES, and it was first solved by Newton. If the bodies do not remain together but eventually separate, their distance apart increasing without bound, then they move in parabolas or hyperbolas—hairpin-shaped curves with ends extending indefinitely.

(Circles, ellipses, parabolas, and hyperbolas were studied by the Greeks two thousand years ago. They are called "conic sections," for they may be obtained by cutting a cone with a plane surface.) In the sky we see many pairs of stars revolving about each other. Observations of these binary systems reveal that they travel in ellipses, so that our confidence in the universality of the Law of Gravitation is strengthened.

One of the famous mathematical problems in celestial mechanics is the problem of three bodies: Given three homogeneous spheres mutually attracting each other in accordance with the Law of Gravitation, determine their motion. This has been the subject of many voluminous papers, for the problem is one of extreme mathematical complexity. Though particular solutions have been found and, given the position and velocity of the bodies at a particular instant, we can compute their configuration at any later time, nevertheless there remain many aspects of the general problem yet to be developed. The same applies in greater degree to the n-body problem, i.e., any number of bodies. In this category we also place the lunar problem, the derivation of the future course of the moon.

AMONG all the deductions from the Law of Gravitation, it should not be surprising to find a few that seem inconsistent with observations. Such departures from theory have been claimed and, though extremely minute, have received considerable attention. This is quite in accord with scientific procedure, for it was a deviation of less than 8 min. of arc (about one-fourth the angular diameter of the moon) from Ptolemaic and Copernican predictions that led to Kepler's laws. At the present time, the perihelion of Mercury is under investigation. It has been claimed that this point is advancing around the sun at a faster rate than is obtained from Newtonian mechanics. While some scientists are checking and rechecking their calculations, theorists are already postulating the actual existence of this deviation and proposing new theories to account for its magnitude. The Einstein Theory of Relativity—a theory that strives for unification of electromagnetic phenomena and gravitation—implies a motion for the perihelion of Mercury which is precisely the same as our observations attribute, namely, 40 sec. of arc per century. Whether this new induction will replace Newton's celestial mechanics or whether a still greater generalization will be discovered is for the future to decide.

THE story of a science has no end, or at least a *finis* that mortal man will ever read, for science is a cycle of observations, inductions, deductions, observations, and so on without end. True, science begins with observations; but no respectable science consists wholly of observations, for, sooner or later in its development, inductions must be made to bring together apparently heterogeneous phenomena. (And now we disclose the source of the apple story—Newton's Law of Universal Gravitation uncovers the association between two apparently different phenomena, the lunar orbit and the fall of an apple.) Our inductions are followed by deductions; that is, from our generalizations we derive particular conclusions. These we put to the test of observation to determine whether or not our inductions are to be modified or abandoned and new inductions formulated. We hope by this process to correlate a greater and greater body of phenomena and arrive at a better understanding of nature.

Though we should congratulate ourselves "that so great a man has lived for the honor of the human race," we should remember the implication of Newton's own remarks: he felt "only like a boy playing on the seashore" and "finding a smoother pebble or a prettier shell than ordinary, while the great ocean of truth lay all undiscovered" before him.

CHAPTER 6

THE SOLAR SYSTEM

> That very law which moulds a tear
> And bids it trickle from its source,—
> That law preserves the earth a sphere,
> And guides the planets in their course.
> —SAMUEL ROGERS (1765–1855)

NATURALLY, we have no definite knowledge concerning the origin of what is known as the SOLAR SYSTEM; but our speculations lead us back billions of years in the past, to the time when there was no earth to measure days or years and when none of its present planet-companions existed. A star, which we now call the SUN, was wandering among many hundreds of millions of other stars, just as it moves today. Did planets travel about this star—our sun? Any objects that might have traveled about it will not enter into our story, for, in the mêlée we are going to discuss, such satellites were annihilated completely or were hurled through space to shift for themselves.

After tranquilly pursuing its course for probably millions of billions of years, our sun passed rather too close to another star. How close was the encounter is a splendid topic for argument on the part of noted astronomers, who place the minimum distance

THE SOLAR SYSTEM

anywhere from ten million to a thousand million miles. The two stars, one of which you remember was our sun, passed each other on the arcs of hyperbolas; and, just as the moon produces tides on the earth, so the gravitational attraction of the two stars produced "tides" on their surfaces. Now even today we see eruptions from the sun's surface and matter shot off hundreds of thousands of miles into space. We can only guess how great was the agitation of the sun as it neared this other star and how huge were the bolts projected.

Much of the ejected material must have been emitted with a velocity so high that it left the sun never to return—if the sun, at the time of the encounter we speak of, were the same size as it is now, a velocity of only 383 mi. per sec. would have been sufficient. Nowadays, bodies leaving the sun with velocities less than 383 mi. per sec. would almost certainly fall back into it. In fact, the only way they could avoid returning would be by passing close to a planet, the probabilities for which are very small. But we must take into account the dynamical action of the passing star. As this star swept by, its gravitation gave to the ejected material an angular motion about our sun in precisely the same direction as the star was moving.

Its visit lasted only a few years and the star continued on its journey, mingling with crowds of other celestial objects, so that we shall never be able to single out the mischief-maker among the billions of stars; perhaps it, too, was blessed with a family of planets from this encounter. Our concern is with the sun and the material left revolving about it. Much of it fell back into the sun, tending to impart a rotational motion in the same direction as the passing star. The larger "chunks" of ejected matter gradually swept up smaller particles and grew to form the planets.

THERE are a number of versions of this tale, as expounded by twentieth-century astronomers; but most of the stories agree that it was a passing star that assisted in the birth of the planets. Skeptics are justified in demanding the basis for these speculations.

The answer is: Our present-day observations *plus* our inferences from these observations. The problem, however, may be likened to the task of reconstructing the opening chapters of a novel, given a couple of badly torn pages from the middle of the book—any result will be subject to question. Nevertheless, before judging this beginning, we must scrutinize the legible portions.

FIRST, we consider the solar system as a whole. From the standpoint of mass, all objects in the solar system are negligible in comparison with the sun, for its mass is more than seven hundred times the combined mass of all the known planets. Jupiter, the most massive of the planets, with a mass greater than three hundred earths, is less than a thousandth the mass of the sun. The bore, who constantly utters his favorite platitude that the world is a small place, does not know that, in this, at any rate, he is sublimely right, for our earth is really quite an insignificant part of the solar system. Notwithstanding, we naturally regard our earth as rather important and feel justified in describing those other fragments comparable with it.

In the order of their distance from the Sun, the list reads at present: Mercury, Venus, Earth, Mars, Jupiter, Saturn, Uranus, Neptune, and Pluto. Very few fully appreciate the enormous size and extent of the members of the solar system. For this reason we suggest that, instead of continuing to construct those ridiculously small models of the solar family, we exert ourselves a little and erect a model more worthy of the system—say one in which a foot represents 500 mi. A globe slightly less than 16 ft. in diameter would represent the Earth. On such a globe, we could still see man, provided we used a microscope magnifying five thousand times, and then he would appear about a tenth of an inch in height; two billions of very small microbes could represent the human race, and specks of dust would make the mountains. The hardest task would be to depict the Sun, for we would need a sphere a third of a mile in diameter. Let us therefore convert our miniature Sun into an office building of a hundred and seventy stories and place it on Chicago's lake front, to dwarf all other skyscrapers.

THE SOLAR SYSTEM

If we maintained our same scale for distances, then Mercury would be represented by a 6-ft. sphere, stored in a kitchenette apartment in Evanston, Illinois. Venus would be a sphere 5 in. shorter in diameter than the Earth, in Gary, Indiana, while the Earth itself would be in Michigan City, Indiana. A block away

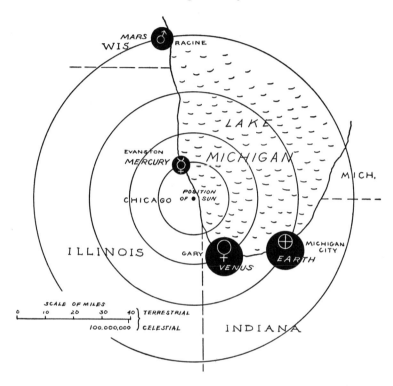

from the Earth—that is, a tenth of a mile distant—would be the Moon, a 4-ft. globe. The $8\frac{1}{2}$-ft. Martian sphere we place in Racine, Wisconsin. The reader will note that the planetary models so far mentioned could all be placed together in a department-store window and, at most, are within about two hours' drive of each other; but it must be remembered that every 2 ft. we travel corresponds to 1,000 mi. Because of their proximity, these planets are called the INNER PLANETS; or, since they more or less resemble our earth in magnitude and density, they are frequently referred to as the TERRESTRIAL PLANETS. The largest and densest is the Earth—in fact it has the greatest average density of all the planets.

Jupiter, the largest planet of all, on this agreed scale would be a sphere more than 170 ft. in diameter, just a trifle flattened at its poles, and might be converted into a unique apartment building in Springfield, Illinois. Saturn would differ in diameter from Jupiter by only 29 ft., but its rings would extend outward from its surface another 100 ft. approximately. Great difficulty would be en-

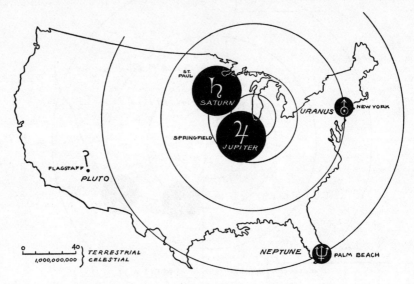

countered in constructing the rings, for, to preserve our scale for them, their thickness would be less than a quarter of an inch. This interesting planet we present to either Minneapolis or St. Paul, Minnesota. The two remaining major planets, Uranus and Neptune, would make modernistic homes with diameters of a little more than 60 ft. and would be placed in New York City and Palm Beach, Florida, respectively. The diameter of the last planet, Pluto, is still subject to question but probably would be only a few feet. Oh, well, we shall place it in the Lowell Observatory at Flagstaff, Arizona, pending final decision.

The purpose of this hypothetical model is merely to indicate relative sizes and distances. The curvature of the earth prevents us from indicating motions with a model on this grand scale, though there would be no difficulty in moving the sphere representing Mercury, the swiftest of the planets, corresponding to its average velocity of 30 mi. per sec. about the sun; Pluto, on the other hand, would travel only a tenth as fast. All the planets journey in the same direction about the sun and, excluding Mercury and Pluto, their paths are almost circular and in the same plane.

	Mercury ☿	Venus ♀	Earth ⊕	Mars ♂	Jupiter ♃	Saturn ♄	Uranus ⛢	Neptune ♆	Pluto
Mean distance from sun in millions of miles	36	67	93	141	489	886	1,782	2,793	3,670
Variation in distance (aphelion − perihelion) in millions of miles	15	0.9	3	26	47	103	168	48	1,800
Inclination of orbit to ecliptic	7°	3°.4	0°	1°.9	1°.3	2°.5	0°.8	1°.8	17°.1
Period of revolution in years	0.241	0.615	1.00	1.88	11.86	29.5	84.0	164.8	248.0
Duration as evening or morning star (days)	58	292	......	390	200	189	185	184	184
Mean velocity in miles per second	29.7	21.7	18.5	15.0	8.1	6.0	4.2	3.4	2.9
Radiation received from sun per unit area (Earth = 1)	6.9	1.9	1.00	0.43	0.04	0.01	0.003	0.001	0.0006
Magnitude at brightest	−1.2	−4.1	......	−2.8	−2.5	−0.2	5.7	7.6	14.0
Diameter in miles	3,000	7,600	7,900	4,200	87,000	72,000	31,000	33,000	?
Mass (Earth = 1)	0.04 ?	0.8 ?	1.00	0.11	317.0	95.0	15.0	17.0	0.1 ?
Average density (water = 1)	3.8 ?	5.1 ?	5.5	4.0	1.3	0.7	1.4	1.3	?
Surface gravity (Earth = 1)	0.3 ?	0.9 ?	1.00	0.4	2.6	1.2	1.00	1.00	?
Period of rotation	88 days?	?	23^h56^m	24^h37^m	9^h50^m to 9^h55^m	10^h14^m to 10^h38^m	10^h40^m	15^h ?	?
Number of known satellites	0	0	1	2	9	9	4	1	0

TABLE 6

STATISTICS OF THE PLANETS

Our model is not complete. There are the satellites of planets other than the Earth, as well as planetoids, comets, and meteors to be added. Nevertheless, we have the framework of the system; and we can easily add, as required, these objects of much lesser mass.

For those readers who enjoy statistics, the accompanying table of physical dimensions (Table 6) is given for comparison. Three of the quantities listed in Table 6—mean distance, variation in distance, and inclination of orbit—partly define the path pursued by the planet. Three additional quantities, however, are needed in order to locate the planet at any specified time, viz., the point where the object crosses the ecliptic from south to north (i.e., the ascending node), the location of the perihelion (i.e., the point where the planet is nearest the sun), and the time of a perihelion passage. But the determination of the planet's position from these six numbers, or ELEMENTS OF THE ORBIT, is by no means a simple task, unless we resort to graphical methods.

CHARTS 5 and 6, in conjunction with Table 7, enable us to ascertain quickly the approximate location of Mercury, Venus, Mars, Jupiter, and Saturn at any time in the interval 1400–2400 A.D. In Chart 5 the orbits of the terrestrial planets, indicated by their symbols, are drawn to scale but, by necessity, in the same plane. A wire model showing the inclination of their paths, though superior to Chart 5, is obviously out of the question in a book. We have, however, indicated by a heavier arc that portion of each orbit lying north of the ecliptic. Chart 6 indicates the corresponding orbits of Jupiter and Saturn, along with that of the earth, but on a different scale.

In both charts, the monthly position of the Earth is indicated while decimals 0, 0.1, 0.2, etc., up to 0.9 designate orbital positions of the remaining planets, the numbers 0 and 0.5 defining perihelion and aphelion, respectively. The order of the decimals indicates the direction of motion, and the arc between the points associated with two successive decimals is the path described by the planet in one-tenth its period of revolution; consequently, given the length of time that has elapsed since a perihelion passage, it is possible by means of these markings to locate the object in either Chart 5 or 6.

The decimal corresponding to the planet's position at any specified time is obtained from Table 7 by a simple addition of numbers associated with the day of the month, the month, and the units, tens, and hundreds in the year number. An example illustrates the method:

We seek the position of the planet Venus on February 5, 1934: The day of the month is 5, the month February, the units in the year number are 4, the tens

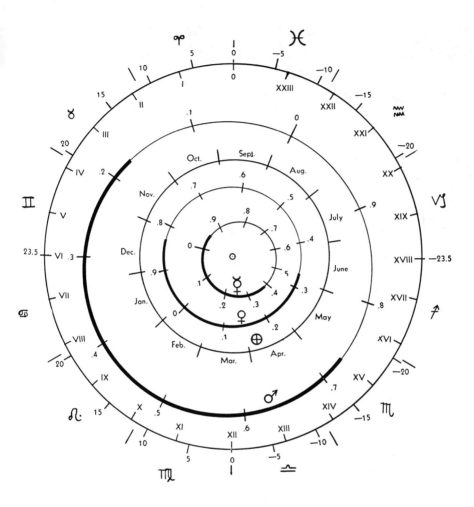

CHART 5

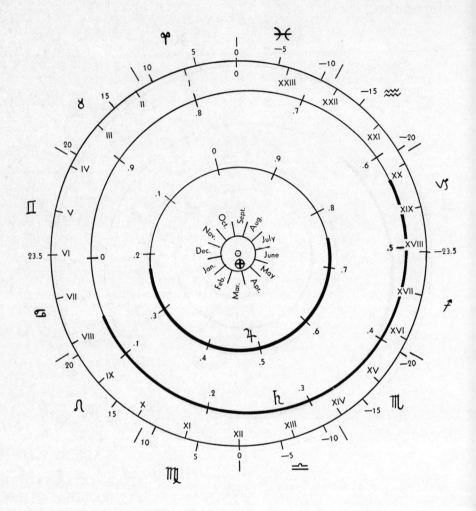

CHART 6

THE SOLAR SYSTEM

Day of Month (Gregorian Calendar)

Day of month..	1	4	7	10	13	16	19	22	25	28	31
Mercury..	0.00	0.03	0.07	0.10	0.14	0.17	0.20	0.23	0.27	0.31	0.34
Venus....	.00	.01	.03	.04	.05	.07	.08	.09	.11	.12	.13
Mars.....	0.00	0.00	0.01	0.01	0.02	0.02	0.03	0.03	0.04	0.04	0.04

Month

	Jan.	Feb.	Mar.	Apr.	May	June	July	Aug.	Sept.	Oct.	Nov.	Dec.
Mercury..	0.29	0.64	0.96	0.31	0.65	0.00	0.35	0.70	0.06	0.40	0.75	0.09
Venus....	.58	.72	.85	.99	.12	.26	.39	.53	.67	.80	.94	.07
Mars.....	.90	.95	.99	.03	.07	.11	.15	.20	.25	.29	.34	.38
Jupiter...	.61	.62	.62	.63	.64	.65	.65	.66	.66	.67	.68	.68
Saturn...	0.49	0.49	0.49	0.50	0.50	0.50	0.50	0.51	0.51	0.51	0.52	0.52

Year—Units

	0	1	2	3	4	5	6	7	8	9
Mercury..	0.00	0.15	0.30	0.46	0.61	0.76	0.91	0.06	0.22	0.37
Venus....	.00	.63	.25	.88	.50	.13	.75	.38	.00	.63
Mars.....	.00	.53	.06	.59	.13	.66	.19	.72	.25	.78
Jupiter...	.00	.08	.17	.25	.34	.42	.51	.59	.67	.76
Saturn...	0.00	0.03	0.07	0.10	0.14	0.17	0.20	0.24	0.27	0.31

Year—Tens

	0	1	2	3	4	5	6	7	8	9
Mercury..	0.00	0.52	0.04	0.56	0.08	0.60	0.12	0.64	0.16	0.68
Venus....	.00	.25	.51	.76	.01	.27	.53	.78	.04	.29
Mars.....	.00	.31	.63	.95	.27	.58	.90	.21	.53	.85
Jupiter...	.00	.84	.69	.53	.37	.22	.06	.90	.74	.59
Saturn...	0.00	0.33	0.68	0.02	0.35	0.70	0.04	0.38	0.72	0.06

Year—Hundreds

	14	15	16	17	18	19	20	21	22	23
Mercury..	0.02	0.22	0.41	0.61	0.80	0.00	0.20	0.39	0.59	0.78
Venus....	.29	.82	.36	.91	.45	.00	.55	.09	.64	.18
Mars.....	.18	.34	.51	.67	.84	.00	.16	.33	.49	.66
Jupiter...	.85	.28	.71	.14	.57	.00	.43	.86	.29	.72
Saturn...	0.02	0.42	0.82	0.21	0.61	0.00	0.39	0.79	0.18	0.58

TABLE 7

"Decimals" for Planets

are 3, and the hundreds are 19. From Table 7 we obtain certain numbers associated with the aforementioned, which we tabulate and sum as follows:

		Associated Number
Day of the month	5	0.01
Month	Feb.	.72
Units in year number	4	.50
Tens in year number	3	.76
Hundreds in year number	19	0.00
Total		1.99

We disregard the number to the left of the decimal point and retain only the decimal part, i.e., 0.99. The point corresponding to 0.99 on the orbit of Venus in Chart 5 represents the position of this planet on February 5, 1934, while the date itself determines the location of the Earth. It so happens that these two points, along with the central point (the Sun), are in a straight line in the order: Sun–Venus–Earth.

Suppose next the date is November 18, 1934. From Table 7 we find for Venus:

		Associated Number
Day of month	18	0.08
Month	Nov.	.94
Units in year number	4	.50
Tens in year number	3	.76
Hundreds in year number	19	0.00
Total		2.28

We place Venus on the point corresponding to the decimal 0.28 and the Earth on a position corresponding to November 18, to discover that these two planets are again in a line with the sun but now the order is Venus–Sun–Earth.

WHENEVER the points corresponding to a planet (other than the Earth) and the Sun fall on a line terminating at the Earth in Chart 5, we say that the planet and the Sun are IN CONJUNCTION. Both examples illustrate conjunctions of Venus and the Sun; and to distinguish between the two different orders, we call the first example an INFERIOR CONJUNCTION (symbolically, "☌ ♀ ☉ Inferior," the symbol ☌ denoting conjunction), and the second, a SUPERIOR CONJUNCTION (☌ ♀ ☉ Superior).

THE SOLAR SYSTEM

In analogous fashion we find for Mercury on March 6, 1934:

		Associated Number
Day of month............	6	0.06
Month	Mar.	.96
Units in year number.....	4	.61
Tens in year number......	3	.56
Hundreds in year number..	19	0.00
Total...................		2.19

On placing Mercury at 0.19 and the Earth at March 6 in Chart 5, we note that these two planets are in a line with the Sun. The conjunction is seen to have been inferior (♂ ☿ ☉ Inferior), for Mercury lay between the Earth and the Sun. Similarly, Chart 5 *plus* Table 7 informs us that on January 20, 1934, Mercury was at superior conjunction (♂ ☿ ☉ Superior).

THOUGH the points corresponding to Earth, Sun, and planet are in a line on the chart, it does not follow that they are actually in a straight line in space. In the first example the "decimal" is 0.99 and the corresponding point is on the *heavier* arc representing the orbit of Venus in Chart 5; consequently, at the inferior conjunction of February 5, 1934, this planet is *north* of the ecliptic and its spacial position is not on a line joining the Earth and the Sun. As seen from the earth, the planet was north of the sun (by about 4°) on this date. Likewise in the remaining examples, the planet involved is either north or south of the sun.

It may happen, however, that at a solar conjunction the planet is on or very close to the ecliptic, and the Sun, planet, and Earth are essentially in a line in space. If this occurs at an inferior conjunction, we on earth will see the planet pass across (i.e., TRANSIT) the sun's disk.

To illustrate, we consider the positions of Mercury and Earth for November 8, 1927:

		Associated Number
Day of month............	8	0.08
Month...............	Nov.	.75
Units in year number.....	7	.06
Tens in year number......	2	.04
Hundreds in year number..	19	0.00
Total...................		0.93

Upon locating the objects on Chart 5, we find that they are in a stright line in the order: Sun–Mercury–Earth; i.e., Mercury is at inferior conjunction with the Sun (♂ ☿ ☉ Inferior). Furthermore, the planet is right at the junction of the light and dark halves of its charted orbit, and it must have crossed the ecliptic from south to north on this date; in astronomical terminology, we say that Mercury was at its ASCENDING NODE (symbol ☊). We are therefore not surprised to learn that Mercury was seen to transit the sun's disk on November 8, 1927.

THE next three transits of Mercury will occur on November 12, 1940; November 13, 1953; and May 5, 1957. As before, we find that for the May transit the planet is at the junction of the light and dark halves of its charted orbit but that it will cross the

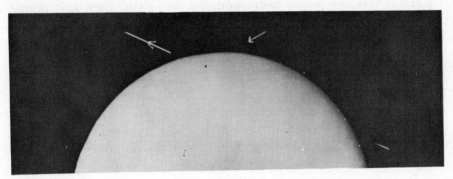

PLATE 6 Transit of Mercury over Sun's Disk, November 14, 1907

ecliptic from north to south, i.e., the conjunction will occur at DESCENDING NODE (symbol ☋). On May 11, 1937, there will be a partial transit of the sun, Mercury apparently grazing the sun.

Unfortunately, the transits of Mercury cannot be seen without suitable optical aid; and we hasten to warn the reader not to look at the sun directly with a telescope, for his eyesight might be seriously damaged; he should use his telescope to cast an image of the sun on a screen.

The student may test his ability in manipulating Chart 5 with the dates of the transits of Mercury during the twentieth century:

THE SOLAR SYSTEM
175

Transit at Ascending Node	Transit at Descending Node	Transit at Ascending Node	Transit at Descending Node
November 14, 1907		November 6, 1960	May 9, 1970
November 6, 1914	May 7, 1924	November 9, 1973	
November 8, 1927		November 12, 1986	
November 12, 1940		November 14, 1999	
November 13, 1953	May 5, 1957		

Careful study of the chart will reveal why, for many centuries to come, the transits occur on about the same date and how Mercury's variation in distance from the sun causes the November transits to be more frequent.

TRANSITS of Venus across the sun are much rarer than the transits of Mercury. The first recorded transit of Venus was predicted by Jeremiah Horrocks, a curate of Liverpool, England, for November 24, 1639 (Old Style). We shall verify his prediction. November 24, 1639 (O.S.) we change to the corresponding Gregorian date, December 4, 1639 (N.S.). On consulting Table 7, we find:

		Associated Number
Day of month	5	0.01
Month	Dec.	.07
Units in year number	9	.63
Tens in year number	3	.76
Hundreds in year number	16	0.36
Total		1.83

We locate our points on the chart and see that Venus is at its ascending node—the Sun, Venus, and the Earth being essentially in a straight line in the order named. Moreover, the perpetual calendar given in chapter 2 shows that December 4, 1639 (N.S.)—November 24, 1639 (O.S.)—was a Sunday; and history tells us how his reverence conscientiously conducted the church services while the planet made its first contact with the sun, and how the twenty-two-year-old curate rushed to his darkroom as soon as he conveniently could to see, projected upon a screen by his two-shilling telescope, the disk of Venus already on the image of the sun!

The December (ascending node) and June (descending node) transits of Venus that have occurred and will occur between 1600 and 2200 are:

Transit at Ascending Node	Transit at Descending Node	Transit at Ascending Node	Transit at Descending Node
December 6, 1631		December 6, 1882	June 7, 2004
December 4, 1639	June 5, 1761		June 5, 2012
	June 3, 1769	December 10, 2117	
December 8, 1874		December 8, 2125	

INASMUCH as the orbits of the remaining planets—Mars, Jupiter, Saturn, etc.—lie wholly outside the path of the earth, they neither transit the Sun nor appear at inferior conjunction. Conjunctions of these planets are *all* superior, and we simply say that the planet is "in conjunction." To illustrate, consider the position of Jupiter on May 8, 1917. The orbital motion in Chart 6 of this planet, as well as of Saturn, is so slight during the course of a month that we may disregard the day of the month; and we find for Jupiter:

		Associated Number
Month................ May		0.64
Units in year number.....	7	.59
Tens in year number......	1	.84
Hundreds in year number..	19	0.00
Total....................		2.07

Upon plotting positions on Chart 6, it is found that Jupiter, the Sun, and the Earth are on a line in the order named (i.e., ♂ ♃ ☉).

On the other hand, an observer on any one of the planets with orbits lying outside that of the Earth would occasionally see the Earth in inferior conjunction with the Sun; i.e., using our chart, a line could be drawn from the Sun through the Earth to the planet in question. However, when such a configuration occurs, we on earth say that the planet is IN OPPOSITION (symbol ☍). As an example, we verify by Chart 5 on January 28, 1931, Mars was in opposition to the Sun (☍ ♂ ☉). On this date, the planet was almost directly opposite the Sun relative to the Earth; and, consequently, it rose at sunset to cross the meridian at midnight.

THE SOLAR SYSTEM

BECAUSE of their proximity to the earth, oppositions are ideal positions for the telescopic study of the exterior planets—Mars, Jupiter, Saturn, Uranus, and Neptune. This is especially so in the case of Mars, and let us therefore focus our attention on this planet. First of all, we compare the opposition of Mars on January 28, 1931, with some conjunction of this planet and the Sun—to be specific, the conjunction of Mars and the Sun on February 28, 1917.

For the opposition, the decimal obtained from Table 7 is 0.42; for the conjunction, 0.02. On plotting the positions of Mars and Earth for both dates, the chart indicates that Mars's GEOCENTRIC DISTANCE (distance from earth's center) at the opposition was almost one-third that of the conjunction. Nor was Mars at its minimum distance from the Earth at the opposition of January 28, 1931. Observationally, this was rather a poor opposition, for the planet was almost at its maximum distance from the Sun (as already pointed out, decimal 0.5 corresponds to aphelion; 0 to perihelion).

As is verifiable by Chart 5, the nearest approaches of Mars to the Earth occur at oppositions on or about August 24, the actual minimum distance in miles being 34,500,000; for the March oppositions, the geocentric distance of Mars is almost twice this minimum (63,000,000 mi.). At conjunctions in the latter part of August, Mars attains its maximum distance of about 250,000,000 mi. from Earth. Inasmuch as Mars, like all the planets, owes its light wholly to the sun, changes in both geocentric and heliocentric distances affect its apparent brilliancy; at conjunction it is of second magnitude, but at a favorable opposition this ruddy planet is forty times as brilliant as its so-called rival, the fiery Antares (from *anti-Ares*).

The most favorable opposition of recent years was on August 22, 1924. As the distance between Mars and the earth was then less than it was or will be at any other opposition in the nineteenth or twentieth centuries, we therefore feel obliged to locate the earth and Mars on Chart 5 for this date. The decimal obtained from Table 7 is 0.99 or almost 0, i.e., perihelion; and the chart reveals how very close was this opposition. A similar opposition, though at a slightly greater distance, will take place in 1939, for favorable (i.e., August) oppositions occur at intervals of 15 or 17 yr.

Incidentally, this last result is deducible from Table 7, for the numbers in the "units" and "tens" tables may also be regarded as indicating the change in the decimal number. Thus, 0.66 is the change in the Mars "decimal" in exactly 5 yr.; 0.31, the change in 10 yr.;

and 0.66+0.31=0.97, the change in 15 yr.; similarly, the change in 17 yr. is 0.97+0.06= 1.03, or 0.03. Now, adding either 0.97 or 0.03 to any "decimal" increases or decreases it by only 0.03, so that, after 15 or 17 yr., both the Earth and Mars will have about the same orbital positions on identical dates. Furthermore, with greater numbers of years, the change in the "decimal" may be still smaller; and, by trying out 1, 2, 3, etc., yr. in turn, computing for each the corresponding alteration and rejecting all intervals for which the change is greater than that for the smaller periods, we obtain in the case of Mars:

	Interval	2	15	17	32	47	79
	Change in decimal	0.06	0.97	0.03	0.01	0.99	0.99

Similarly, for Jupiter:

	Interval	1	11	12	83	95
	Change in decimal	0.08	0.92	0.01	0.99	0.01

The student may verify that 190 yr. will pass before the Earth, Mars, and Jupiter approximately repeat their maneuvres about the sun. As to the orbital arrangement of all the planets, we may safely say that it is never twice the same.

Chart 6 reveals that the percentage variation in geocentric distance at opposition is of no great consequence in the case of Jupiter and Saturn. The changes in brightness for the major planets, therefore, are not at all comparable with those for Mars; Jupiter's magnitude at minimum geocentric distance is -2.5, at maximum -1.4, the ratio in brightness being about $3:1$; the inclination of the rings of Saturn (to be discussed later) are chiefly responsible for its greater range of magnitudes, -0.4, to 1.2.

THE planets interior to the Earth's orbit—Mercury and Venus —are not at their maximum brilliancy when closest to the Earth, i.e., at inferior conjunction. In our study of the solar system, we must always remember that the source of illumination is the sun and that half of every planet is in sunlight and half in darkness. At the opposition of an exterior planet the illuminated half is directed toward the earth; while, at an inferior conjunction, the planet Venus (and the same applies to Mercury) presents its dark half to us in a manner analogous to the new moon. In fact, as the interior planets Mercury and Venus pass from inferior conjunction to superior and then back to inferior, they exhibit through a telescope all the phases of the moon—phenomena predicted by Copernicus and verified observationally for Venus by Galileo

Inferior — phase
Superior — no phases

(chap. 5). The phase of exterior planets is "full" both at conjunction and opposition, and only by transporting ourselves beyond their orbits could we hope to view them as crescents. Relative to Jupiter and Saturn, we on earth are so close to our common source of light, the sun, that the disks of these planets *always* appear completely illuminated; Mars, however, at minimum phase is distinctly

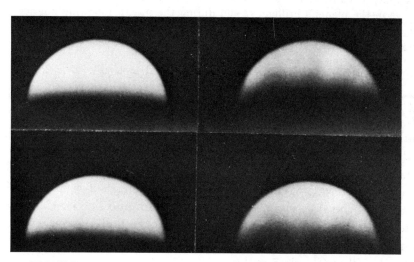

PLATE 7 VENUS AT QUADRATURE, IN VIOLET AND RED LIGHT

gibbous, corresponding to the appearance of the moon when 3 days from full.

Owing primarily to changes in phase and geocentric distance, Venus has the greatest variation of all the planets in apparent brightness. With the aid of Chart 5 we follow the motion of this planet and the Earth from one inferior conjunction to the next.

As a typical example we start with the inferior conjunction of February 5, 1934 (decimal for Venus is then 0.99). Its apparent proximity to the Sun prohibits its detection with the eye and makes telescopic study difficult, though Venus has been observed when less than 2° from the Sun. At inferior conjunction, the planet is only 26,000,000 mi. from the Earth (18,500,000 mi. less than the minimum geocentric distance for Mars); and consequently, its apparent angular diameter attains its maximum—a little more than 1 min. of arc. We relocate Venus about a month later—to be specific, on March 6, 1934 (decimal 0.13). The Sun, Venus, and Earth are no longer in a line; but on our chart the angle between the lines joining the Sun and Venus to the Earth is almost 40°. So in the sky; the planet has increased its angular distance from the Sun from practically zero at inferior conjunction to about 40° in less than 30 days. If, during the interim, we had watched the planet through a telescope, we would have seen it increase from a thin crescent to a phase corresponding to that of the moon when it is 5 days old. Venus' geocentric distance has also increased, and its apparent diameter is a third less. This change in phase produces a decided change in the planet's brilliancy; and during the month following inferior conjunction Venus rapidly brightens, to become the most brilliant point of light in the sky. On March 6, 1934, the magnitude of this planet was -4.1, and it was then plainly visible in the daytime!

THOSE who notice Venus in the sky along with the sun for the first time are often surprised that so conspicuous an object could have escaped their attention in the past. There is little difficulty in finding this planet in the daytime when it is almost at maximum brightness, provided its approximate position is known. In chapter 3 we provided the reader with charts for determining location in the sky, given the right ascension and declination of the object, and now we shall fully equip him for naked-eye observations of the planets by divulging how to derive these equatorial co-ordinates from Charts 5 and 6.

Both charts are bounded by a circle, upon which are placed the signs of the zodiac, Roman numbers 0, I, II, up to XXIII and Arabic numerals 0 to 23.5, denoting, respectively, right ascension and declination. A straight line drawn from the Earth's position to the Sun on Chart 5 for any date, when extended until it cuts the boundary of the chart, indicates on this circle the equatorial co-ordinates of the Sun. (It is unnecessary to mark the chart; "drawing a line" may be accomplished just as well by placing a ruler or any other straight edge upon the chart.) For example, on March 6, 1934 (here the line from Earth to Sun intersects the outer circle at XXIII), we find from Chart 5 that the right ascension of the Sun is 23 hr., declination $-6°$, and, incidentally, the Sun is in Pisces (not the constellation but the *sign* [see chap. 3]). To determine these same quantities for Venus, draw a line from the Earth to the planet (the latter being given

THE SOLAR SYSTEM

in this case by the decimal 0.13), extending the line until it cuts the boundary. This intersection, however, does *not* give the equatorial co-ordinates of Venus. The next step is to start with the Sun and draw a *second* line *parallel* to the first and in the same direction (i.e., from Earth to Venus). Where this *second* line cuts the outer circle we read the right ascension and sign for Venus, viz., right ascension 20^h30^m, sign Capricornus.

Though we stated the right ascension for Venus on March 6, 1934, as derived from Chart 5, we purposely omitted the corresponding declination indicated by the outer circle, viz., declination $-19°$. This would be the declination of the planet, provided it were on the ecliptic; but decimal 0.13 for the date under consideration is on the heavier arc, and Venus therefore was north of the ecliptic. The error introduced by taking the declination of Venus as that given by the outer circle is really only a few degrees; but, for those who insist upon greater refinement, we give here a tabulated list of corrections. The tabular cor-

Decimal	Jan.	Feb.	Mar.	Apr.	May	June	July	Aug.	Sept.	Oct.	Nov.	Dec.
0–0.09	5°	7°	6°	4°	2°	2°	1°	1°	1°	1°	2°	2°
0.10– .19	2	4	7	7	3	2	2	1	1	1	1	2
.20– .29	1	2	4	4	3	2	1	1	1	1	0	1
.30– .39	0	0	0	0	0	0	0	0	0	0	0	0
.40– .49	−1	−1	0	−1	−1	−2	−3	−4	−2	−2	0	−1
.50– .59	−1	−1	−1	−1	−2	−2	−5	−7	−6	−4	−2	−2
.60– .69	−2	−1	−1	−1	−1	−2	−2	−4	−7	−7	−3	−2
.70– .79	−1	−1	−1	−1	0	−1	−1	−2	−4	−4	−3	−2
.80– .89	0	0	0	0	0	0	0	0	0	0	0	0
0.90–0.99	3	4	2	2	0	1	1	1	0	1	1	2

AMOUNT TO BE ADDED TO CHART 5 DECLINATION FOR VENUS

rection for March, decimal 0.13, is 7°, so that the corrected declination is $-19+7 = -12°$. This co-ordinate, in conjunction with the right ascension 20^h30^m, enables us to use the methods of chapter 3 to locate the planet's position at any hour of the day for a specified latitude.

Since the right ascension of the sun on March 6 is 23 hr., Venus crossed our meridian $23-20:30$, or 2^h30^m, *before* the sun, for right ascensions are merely the sidereal time-table of meridional transit. To have seen Venus in all its splendor among the stars on this date, we too must have risen *before* the sun. As was true in this example (the same may be said also of Mercury), whenever Venus disappears while passing the sun at inferior conjunction, the planet reappears to the *west* and heralds the sun. Venus then rises after midnight and is said to be a morning star.

We again locate Venus and Earth, this time for April 16, 1934 (decimal 0.32). The angle between the lines joining the Sun and Venus to the Earth is now 46°; furthermore, the line connecting the Earth and Venus is tangent to (i.e., just touches) the orbit of Venus; and consequently the angular distance between the Sun and Venus attained its maximum (46°) on the date mentioned and the planet was said to be at GREATEST WESTERN ELONGATION. The equatorial co-ordinates of the Sun and Venus were then as follows:

	Sun	Venus
Right ascension	1^h36^m	22^h40^m
Declination	10°	−9°

The planet crossed the meridian $1:36+24-22:40=2^h56^m$, or approximately 3 hr., before the sun. The student may verify these results with Chart 5 in conjunction with the table of corrections for Venus.

In Chart 5 we find the planet, Earth, and the Sun forming a right-angle triangle with Venus at the vertex of the right angle—a configuration analogous to the one the moon forms with the earth and sun when it is a week old; consequently, at greatest western elongation Venus appears to be at first quarter when viewed through a telescope. With this increase in phase, the planet might be expected to continue to brighten; but the increase in distance from March 6, 1934, to April 16, 1934 (an interval of 41 days), offsets this effect, and the planet actually fades a trifle.

After attaining its maximum western elongation, the planet's angular distance from the Sun, as well as its brightness, slowly diminished; and Venus disappeared in the brilliant light of day as it approached its superior conjunction of November 18, 1934. Its phase was then full, though its apparent angular diameter was only 11 sec. (one-sixth that at inferior conjunction) and its brightness four-tenths its maximum brilliancy. At superior conjunction, the planet crosses the Sun from west to east to become an evening star. If he wishes, the student may retrace the course of this planet and the Earth, when Venus slowly approached its GREATEST EASTERN ELONGATION on June 30, 1935 (telescopic appearance similar to the moon at quarter), attained maximum brilliancy as an evening star on August 13, 1935, and then hastened to pass the Sun on September 8, 1935, to repeat the cycle.

To summarize, we list in order for Venus the different phenomena from one inferior conjunction to the next:

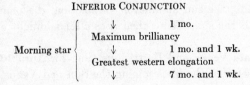

THE SOLAR SYSTEM

Evening star
{
SUPERIOR CONJUNCTION
↓   7 mo. and 1 wk.
Greatest eastern elongation
↓   1 mo. and 1 wk.
Maximum brilliancy
↓   1 mo.
}

INFERIOR CONJUNCTION

The intervals are only approximate and may vary by a few days, but the complete cycle averages 584 days (about 19 mo. and 1 wk.) and is called the SYNODIC PERIOD of Venus.

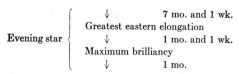

Venus 46°
Mercury 18°–28°

MERCURY likewise exhibits all the phases of the moon, but with varying apparent angular size, and has a cycle of positions similar to those of Venus. Though the angular distance of Venus from the sun at greatest eastern or western elongation remains always around 46°, that for Mercury varies from 18° to 28° because of the variation in this planet's heliocentric distance. Mercury is best seen in the evening at a time of greatest eastern elongation, particularly when this occurs in April or March, for then the planet is at its minimum heliocentric distance and maximum brilliancy (magnitude −1.2; see Chart 5); on the other hand, as a morning star we should look for it when it happens to be at greatest western elongation, in September and October. Because of its angular proximity to the sun, we must search for it in the twilight just after sunset or before sunrise.

The method of determining the right ascension and declination of Venus by means of Chart 5 is applicable to all the planets. We could include tables for correcting the declination as read from the outer circles of Charts 5 and 6, but this is unnecessary for approximate position. Only in the case of favorable oppositions of Mars does this departure become appreciable, but then the planet is so very bright that a rough location is quite sufficient for identification. The reader may test his ability to use the charts by checking the positions of the planets and sun for June 11, 1934, as given in the accompanying table. A planet is termed an "evening star" if it is above the horizon at sunset, a "morning star" if above at

	Decimal	Right Ascension	Declination	Sign	Description
Sun		5ʰ15ᵐ	17°	♊	
Mercury	0.28	7 00	24	♋	Evening star
Venus	.56	2 30	13	♉	Morning star
Mars	.20	4 15	22	♊	Morning star
Jupiter	.52	12 50	−4	♎	Evening star
Saturn	0.66	22 00	−13	♒	Morning star

sunrise. The duration of a planet as a morning or evening star is given in Table 6. This is one-half the synodic period, which is the interval from one superior conjunction to the next. The figures just given are the actual values; and the student, therefore, will be unable to duplicate them exactly, especially the column of declinations, for the chart does not take inclination of orbits into account (except for Venus).

As another exercise, he might work out the planetary positions at the time of his birth; and in this connection he ought to read Edmund's opinion of this pursuit in *King Lear*, Act I, scene ii. We hope that the charts not only will enable him to locate the planets but will lead him to a better understanding of how the apparently peculiar courses they pursue among the stars are a consequence of the regular orbital motion of the earth and the planets around the sun.

THE planetary motions in our sky are excellently and faithfully reproduced by that ingenious instrument—the Zeiss Planetarium. The first planetarium was erected in Munich in 1925, and there are now (1935) nineteen such instruments in existence. The Adler Planetarium and Astronomical Museum, Chicago, opened in 1930, brought to the United States not only this intricate device but also a superb collection of antique astronomical instruments. Three years later a second planetarium was opened in this country at the Franklin Institute in Philadelphia. This was followed by one in Los Angeles, and now the Hayden Planetarium of New York City is under construction.

By means of a complex projector, a representation of the entire heavens may be thrown on the interior of a hemispherical dome. Not only may the instrument be set to depict the stellar skies of any latitude and time, but also it may be driven at various speeds

THE SOLAR SYSTEM

to show the apparent celestial motions of the sun, moon, visible planets, and stars in intervals ranging from a day to the Great Year (25,800 yr.). We urge the reader to visit one of these planetaria when he has an opportunity, for they are designed primarily for popular instruction in astronomy.

We shall now consider the members of the solar family individually:

MERCURY

MERCURY, the planet nearest the sun, is the swiftest moving planet and hence is aptly named after the messenger of the gods, the symbol ☿ of the planet being the caduceus or wand of the god Mercury. The variation in distance of the planet from the sun is quite appreciable. Its maximum distance from the sun is 43,000,000 mi., and at this aphelion point Mercury travels with a velocity of 23 mi. per sec. Forty-four days later it approaches perihelion, its position of minimum distance from the sun, which is about 29,000,000 mi., and speeds up to its maximum velocity of 36 mi. per sec. Mercury then slows down until it passes aphelion, another 44 days later. If the messenger of the gods traveled as fast as his namesake, he could have made a transatlantic flight in less than 2 min.

When the sky is very clear and we have an uninterrupted view of the horizon, we may sometimes see Mercury, just after sunset or just before sunrise, as a bright sparkling star right in the sun's twilight. Despite the fact that, at maximum brilliancy, it is almost as bright as Sirius, Mercury is difficult to detect because of its proximity to the sun. Even Copernicus lamented in his later years that, try as he would, he had never seen this planet. An old English writer humorously terms it "a squinting lacquey of the sun, who seldom shows his head in these parts, as if he were in debt."

The motion of Mercury from one side of the sun to the other puzzled the ancients, and they failed to recognize its "dual personality." As a morning star it was called Apollo, the god of day; but, when it appeared in the evening, it was Mercury, the god of thieves prowling in the night seeking to plunder and despoil. Poor Mercury, what a variety of names—"The Sparkling One" of the Greeks; "The Malignant Planet" of the astrologers.

Telescopic observations of this planet are rather disappointing. It is best studied during the daytime, for, when it is visible at night, it is close to the horizon, and thus its image becomes distorted by the thick layer of atmosphere through which its light must pass. No definite markings have as yet been detected on the planet either with the naked eye or with photographic plates. But we cannot conclude that no markings exist, for, if Mercury were replaced by our moon, the only way the substitution would be detected would probably be by the 30 per cent decrease in apparent size. As with the moon, no atmosphere has been detected on the planet, for, as Mercury transits the sun, it appears merely as a black dot on the sun's disk.

What are the climatic conditions on this planet? Lacking definite markings, we cannot watch it rotate, and no evidence of a rapid rotation has been obtained so far; hence some observers conclude that the planet always presents the same face to the sun, as the moon presents the same face toward the earth. But, just as we are able to see a little beyond the east or west limb of the moon because it rotates uniformly and revolves at a variable angular rate, so may we conclude that, even if Mercury's period of rotation and revolution were equal, more than half of its surface would be subjected to the sun's broiling rays. Whether or not the planet rotates, we conclude that, because of its proximity to the sun, the temperature of the sunlit portion is so high—and there are observations to justify this conclusion—that the Mercurian lakes could well be of molten lead.

VENUS

VENUS is called the sister-planet to the earth, for, of all the planets, it is the one which most closely resembles the earth. In diameter it is only about 200 mi. less than the earth, and its distance from the sun is around 25,000,000 mi. less. Its orbit is the most circular of the planets, and it revolves about the sun every 225 days.

Named after the Queen of Beauty and having ♀ (the looking glass) for its symbol, the planet reflects 60 per cent of the light that falls on it from the sun—about ten times the corresponding percentages for Mercury and our moon. Through a telescope the planet appears spotlessly and intensely white, and we conclude that it is covered with a dense layer of white clouds. That Venus has a very dense atmosphere is seen when the planet appears as a thin crescent. Instead of the ends terminating on a diameter, as was the case with the moon, they extend far beyond and sometimes even meet, indicating that this continuation of the crescent is due to sunlight illuminating the planet's atmosphere. Then, again, when Venus is seen to cross in front of the sun's disk, the planet is surrounded by a halo of light.

As was the case with Mercury, the absence of any definite markings on the disk of Venus, as seen through a telescope, has prevented us from obtaining reliable information about its period of rotation. Perhaps a further development of photography, with non-visible radiations that penetrate the dense layer of Venus' clouds, will reveal markings more definite than the few hazy shadows which have been obtained. Using an instrument known as the SPECTROSCOPE,*

---

* Lemon, *From Galileo to Cosmic Rays*, p. 359.

THE SOLAR SYSTEM

it is possible to detect and measure rapid rotations by studying the nature of the light reflected by the planet. This has been done for both Mercury and Venus; and, since no rapid rotation was discovered in either case, the conclusion, though open to question, was that their periods of rotation were greater than a week. Then, again, it has recently been found that some heat is radiated by the dark portion of the planets; and as a result, certain astronomers believe that Mercury and Venus do not always present the same face to the sun.

Since so little is known about the rotation of Venus, any statements about climatic conditions on the surface are mere speculations. Of course, the portion of the planet subjected to the direct rays of the sun will be at a high temperature, and recent measurements of heat radiated place this at about 50° C. (122° F.). On the dark side, similar measurements indicate a temperature of −30° C. (−22° F.). Is there life on the planet Venus? We refer the reader to the Sunday supplements and the tabloids—they seem to have sources of information not available to the astronomer. Indications are that there is little, if any, oxygen (the method of discovering the presence of gases will be discussed later) in the upper atmosphere of Venus, as well as an absence of water vapor. Oxygen being associated with life, some astronomers venture to declare that Venus is like the Sahara Desert; but it might well be that oxygen and water vapor are present in the lower levels of the atmosphere. Besides, we must leave some loophole to justify the newspapers presenting us with those amazing pictures of Venusian men and women, who, living under a thick mantle of clouds, see neither stars nor planets nor sun. To them their Venus is the universe.

THE EARTH

MORE to emphasize that the Earth is a planet with its own symbol, ⊕ (a meridian and equator), we again look at it. But this time let us look at it from Venus—of course from a point above the Venusian clouds. When Venus is at inferior conjunction, we on Venus see the Earth crossing our meridian at midnight, assuming, of course, that Venus rotates. The Earth is then the brightest object in the Venusian sky, excluding, of course, the Sun, and is about twice as bright as Venus appears at its brightest from the Earth. The Moon appears as a typical first-magnitude star, not quite as bright as Sirius, however, oscillating from one side of the Earth to the other, at most a little more than half a degree from it. (Note that half a degree is the angle subtended by the Sun or Moon from the Earth.) Should we call the Moon a satellite of the Earth? On Venus, the Moon is clearly visible to our unaided eye and would appear to be a planet. The Earth and Moon would be the twin planets for, as we shall find, no other planet has a satellite comparable with itself. Telescopically, our Earth would be a more interesting sight to Venus than Venus is to us, for, when closest to Venus, the Earth presents its entire illuminated half.

MARS

PASSING outward from the earth's orbit we next meet the ruddy-colored planet Mars. Named after the god of war, the planet's sign is ♂, the shield and spear.

The telescopic appearance of Mars is far more interesting than that of cloud-covered Venus. The major portion of the planet is reddish, and this accounts for its ruddy appearance to the eye. Certain dark markings found by

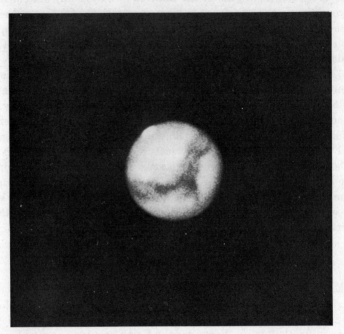

PLATE 8 MARS—REGION OF SYRTIS MAJOR

Cassini in 1666 were so permanent that they are still identifiable, and we conclude that what we see is the solid portion of the planet. Even in a few hours, we can determine from the motion of these markings the Martian day, that is, the rotation period of Mars, which has been found to be $24^h 37^m 22^s.6$ of mean solar time, the direction of rotation being eastward, as with the earth. It seems remarkable that a planet differing so radically in size from the earth (Mars's diameter is only 4,230 mi.) should have a day just 41.5 min. longer. But perhaps more astounding is the fact that the Martian equator is inclined $23\frac{1}{2}°$ to its orbit, the same angle as that between earth's equator and ecliptic! The Martian climatic zones therefore are similar to those of the earth, but the seasons are almost twice as long as ours. Furthermore, the perihelion point of the orbit of

THE SOLAR SYSTEM 189

Mars is so located that, as with the earth, the winters in the northern hemisphere are shorter than the summers. The present Martian north celestial pole, however, is close to the star Deneb, occupying, in fact, almost the same position our north celestial pole will have, because of precession, in the year 9000 A.D. Present Martian star maps could therefore be used by ninety-first-century earthlings.

The season on Mars at the time of any opposition is the season the earth will have in three months. At the August oppositions—the positions of minimum distance from earth and the best time to observe Mars—it is almost the beginning of winter in the northern Martian latitudes; the south pole of the planet is tilted toward the sun and earth, and that is why we know more about the southern hemisphere of Mars than the northern. Inasmuch as we can see so clearly the solid portion of Mars, we conclude that the planet has a very thin atmosphere. That it has some atmosphere is indicated by the percentage of sunlight it reflects (15 per cent of that received), which is about twice the fraction reflected by Mercury or the moon. Other observations tell us that the atmospheric pressure at the Martian surface is about that prevailing at the summits of our highest mountains. Discard the atmosphere below a height of 7 mi. on the earth, including the clouds we find there, and the remaining atmosphere (about 25 per cent the mass of the original) would resemble the Martian atmosphere, though this would extend to greater heights with appreciable density, gravity on Mars being a little more than a third that of the earth. No dense layers of clouds are seen on Mars, though faint patches, possibly clouds of dust and mist, have been seen and photographed. These often appear detached from the planet's disk, thus proving they are not markings on the solid surface. Recent observations indicate that both water vapor and oxygen are present in the Martian atmosphere but to a very much smaller extent than on the earth.

AROUND the poles of the planet are very white areas called polar caps, extending down as far as Martian Lat. 55° N. or S.—about the latitude of Scotland on our globe. The caps vary in size with the Martian seasons, being at their largest in the winter for the particular hemisphere; then, as spring approaches, they diminish rapidly, sometimes disappearing before the end of summer. Perhaps they are a precipitation of water from an atmosphere which again evaporates as the temperature of the atmosphere rises. They seem to project a trifle outward from the surface, and it has been suggested that what we see is a dense white mist hovering over frost-covered poles. White patches seen in midlatitudes, visible for a few hours after the sun's rays first strike them, may be night frost.

The large dark permanent markings, grayish-green in color, were thought by seventeenth-century astronomers to be oceans; the still larger reddish regions, continents. Modern telescopes, however, reveal not only changes in these grayish-green regions with the Martian seasons but also a wealth of finer structure on both so-called continents and oceans. As the north polar cap shrinks in size, the darker regions in its vicinity become more intense and this deepening spreads into the lower latitudes toward the equator, gradually fading with the advent of autumn and winter in the northern hemisphere. A similar transformation takes place in the southern hemisphere, corresponding phenomena occurring half a Martian year later. Because of the presence of oxygen in the atmosphere on Mars—a vital element in the processes of plant life, as we know it—these changes have been attributed to vegetation growing in the marshy regions formed by the melting polar caps. Despite the fact that the mean temperature of Mars is probably below freezing-point (theory and observations so far agree in this conclusion), nevertheless measurements of heat radiated indicate that the equatorial regions have a temperature well above freezing point at noon, though descending below freezing point at night. As far as our observations go, therefore, we have found no condition that would prohibit plant life.

On the whole, the surface of Mars is much smoother than that of the earth or the moon. No mountain ranges or peaks have been detected, though when we reflect that what we are observing is, at its closest, a hundred and fifty times as far from us as the moon, we question such statements that "Mars is as smooth as the Sahara Desert." Objects a tenth of a mile in diameter are just discernible on the moon with our largest telescope; and hence, even under ideal conditions, we would have great difficulty in seeing objects 15 mi. in diameter on Mars.

IN 1877 an Italian astronomer, Schiaparelli, announced that he had seen long and very narrow straight dark streaks running for hundreds and even thousands of miles along the arcs of great circles. These he called *canali*, and thus began the bitter arguments about the canals on Mars. Later, Percival Lowell and his associates mapped off more than four hundred of these so-called "canals" to form an intricate network of multiple intersecting and parallel circular arcs on the Lowell Martian globe. With an imagination only surpassed by writers in the Sunday supplements, but with far superior logic, Lowell starts with these and other observations and concludes that Mars is inhabited. A "high" form of civilization is desperately trying to exist by constructing canals to drain the polar caps; the canali are strips of vegetation on either side of the banks of the canals. Altogether, Lowell's story, *Mars as the Abode of Life*, is highly entertaining and so plausible as almost to be convincing; but, unfortunately, there is wide disagreement about the character of these markings. Some very capable astronomers have failed to see any indication of the markings, and Lowell and his associates seem unique in claiming that they are along the arcs of circles. No doubt there *are* markings just beyond the range of visibility, and it might well be that the eye is apt to combine these faint specks into streaks.

THE SOLAR SYSTEM

REVOLVING eastward about the planet in the plane of its equator are two tiny satellites named Deimos and Phobos, or "Dread" and "Terror," the sons of Mars. The inner one, Phobos, is less than 4,000 mi. from the planet's surface; and, judging from the amount of light it reflects, its diameter is about 10 or 15 mi. At its minimum distance, when overhead at the Martian equator, it would appear to the Martians, if Martians there be, about a third to a half the size the moon appears to us, though we must not conclude Phobos would appear as a miniature moon, for we do not know its precise shape. Traveling with a faster angular speed than the planet rotates (Phobos' period of revolution is only 7^h39^m), it appears to move eastward in the Martian sky; and at the equator it rises in the west and sets in the east four and a fraction hours later, to appear again in the west after a total of 11 hr. have elapsed. In a Martian day it would repeat all its phases at least three times. Deimos, about half the size of Phobos, is 10,000 mi. from Mars's surface and would appear in the Martian sky as Venus does to us. Its period of revolution being 31 hr., it is seen to rise in the east and set in the west $2\frac{1}{2}$ Martian days later, passing through all its phases twice in this interval.

In connection with these satellites, we cannot resist the temptation to quote Dean Swift's strangely prophetic description of astronomy in his "A Voyage to Laputa," chapter iii of *Gulliver's Travels:*

They [the astronomers] spend the greatest part of their lives in observing the celestial bodies, which they do by the assistance of glasses, far excelling ours in goodness. For, although their largest telescopes do not exceed three feet, they magnify much more than those of a hundred with us, and show the stars with greater clearness. This advantage has enabled them to extend their discoveries much farther than our astronomers in Europe; for they have made a catalogue of ten thousand fixed stars, whereas the largest of ours does not contain above one-third part of that number. They have likewise discovered two lesser stars, or satellites, which revolve about Mars; whereof the innermost is distant from the centre of the primary planet exactly three of his diameters, and the outermost, five; the former revolves in the space of ten hours, and the latter in twenty-one and a-half; so that the squares of their periodical times are very nearly in the same proportion with the cubes of their distance from the centre of Mars; which evidently shows them to be governed by the same law of gravitation that influences the other heavenly bodies.

They have observed ninety-three different comets, and settled their periods with great exactness. If this be true (and they affirm it with great confidence), it is much to be wished that their observations were made public, whereby the theory of comets, which at present is very lame and defective, might be brought to the same perfection as other parts of astronomy.

Now, *Gulliver's Travels* appeared in 1726, while Mars's two tiny satellites were discovered by Asaph Hall at Washington in 1877!

If a planet has satellites, it is a simple task to determine its mass (see chap. 5). We compare the accelerations the particular planet gives to its satellite, in compelling the latter to revolve about it, with the acceleration the earth gives to the moon. The mass of Mars is thus found to be one-tenth that of the earth. The same method is used for the determination of the sun's mass (332,000 times that of earth), the "satellite" in this case being the earth. The mass of either Mercury or Venus is more difficult to determine, and the results are not so reliable. The method in the case of these planets is to measure the slight perturbations produced by them on other planets and on our moon.

Knowing the mass and radius of Mars, we find that its surface gravity is a little more than one-third that of the earth; the Martian canal-diggers could handle shovels with three times the capacity of ours, assuming the same physical strength in both cases. Then, again, by being projected with a velocity of 3 mi. per sec., the Martians could escape the gravitation of their planet. What a simple matter it would be for the Lowell- and newspaper-inhabitants of Mars to shoot themselves off with a velocity of $2\frac{1}{2}$ mi. per sec. for a week-end on Phobos. And what an exciting time they would have there! How light-headed they would feel as they tried to keep their feet on the Phobosean ground, for there they would weigh only an ounce or so; a little too lively a step would start them slowly soaring hundreds of feet. They would also be compelled to drive their cars in low gear and very cautiously, for, if they traveled too quickly up a hill, they might leave Phobos, to fall to their destruction on Mars.

JUPITER

MORE than five times as far from the sun as the earth is the largest of the planets, Jupiter, named after the king of the gods and represented by the sign ♃, a hieroglyphic symbolizing an eagle, "the bird of Jove." Slowly and majestically we see it traveling among the stars, each year moving from one sign of the zodiac to the next, completing its circuit among the stars in a little less than 12 yr.

When Galileo pointed his telescope at the king of the planets in the beginning of the seventeenth century, he saw a miniature solar system within the solar system—a planet about which revolved four satellites. These four satellites are almost visible to the unaided eye; in fact, one Russian traveler stated that a hunter in Siberia pointed to Jupiter and said, "I have just seen that star swallow a small one and then vomit it up again." To these Galilean satellites our powerful telescopes have added five more. Although the latter five are rather small, ranging in size from 10 to 100 mi., the Galilean satellites are comparable with our moon; in fact, two of them are a little greater than Mercury in size. These four satellites are the largest objects in the Jovian sky, the nearest appearing as large as the moon appears to us, the most remote a little more than one-fourth the apparent diameter of the moon.

From Jupiter our sun would have one-fifth the apparent angular diameter it has from the earth; and, inasmuch as these four moons are always near the plane of Jupiter, solar and lunar eclipses are daily occurrences. The first three

Galilean satellites—Io, Europa, and Ganymede—with periods of 1^d18^h5, 3^d13^h, and 7^d4^h, respectively, cast their tiny shadows on the huge planet every time their phase becomes new; and they themselves likewise become eclipsed by the planet when their phase is full. As they cast their shadow on Jupiter, transit its disk, become eclipsed and disappear behind it, these satellites provide the explorer with a celestial clock. The sun and stars form a clock, as has already been pointed out; but, to determine standard time by these objects, one must know one's longitude. The Jovian clock, however, is independent of longitude, for, our earth being so small in comparison with Jupiter's distance, the arrangement of the Galilean satellites will appear the same for all points of the earth. With a low-powered telescope to see the satellites and a copy of the *Almanac* to read the Jovian clock, the explorer is able to determine Greenwich Mean Time. If a sailor is wrecked and his chronometer destroyed, a small telescope, a copy of the *Nautical Almanac*, and a clear sky to see Jupiter will guide him home.

In 1675 the Danish astronomer O. Römer noticed that this Jovian clock appeared to run fast as we approached Jupiter in our motion about the sun, and to run slow as our distance from Jupiter increased. Römer concluded from these phenomena that light has a finite velocity—the light of this clock taking a definite time to reach the earth. As we approach Jupiter, Jovian time reaches us earlier and the clock appears to run faster. We thus owe

PLATE 9 JUPITER, SHOWING CHANGE IN POSITION OF FOUR SATELLITES

the discovery of the finiteness of the velocity of light to astronomical measurements. If Römer had known the size of the earth's orbit, he could have determined the velocity of light and would have found it to be the same value as that now obtained in physics laboratories.

The nearest seven satellites of Jupiter all travel eastward about the planet; but the eighth and ninth, at a distance of 14,600,000 and 15,000,000 mi., travel westward. It is not at all surprising that the last two travel in an opposite direction; but it is surprising that the first seven travel eastward, for it can be shown by the principles of celestial mechanics that, close to the planet, a satellite can travel either eastward or westward, but that at greater distances, because of the sun's attraction, they can travel westward but not eastward, while still further distant, they are pulled away from the planet by the sun, no matter in which direction they travel. It almost seems that, given a choice, a satellite will go eastward rather than westward.

JUST as with Mars, we find the mass of Jupiter, from the motion of its satellites, to be more than three hundred times that of the earth. The volume of Jupiter, however, is thirteen hundred times that of the earth, and hence the average density of the giant planet is considerably less than that of the latter—about one and a third times that of water. It should be noticed that this is an "average density." The density of the interior is, without doubt, much greater than the average; while the exterior, the point which we see through our telescopes, is a thick gaseous atmosphere.

Quite in contrast with the earth's atmosphere, the Jovian "clouds," or rather atmospheric markings, are fairly permanent features. From these atmospheric markings we find that this huge planet rotates eastward faster than the earth, a Jovian day being less than 10 hr. Furthermore, equatorial markings complete their circuit in $9^h 50^m$, while those in higher latitudes require $9^h 57^m$; for every 85 Jovian days at the equator, there are therefore only 84 in the northern latitudes. As might be expected from its high angular speed and size, the planet bulges at the equator; and the oblateness, or flattening at the poles, is quite noticeable in photographs obtained.

THE most noticeable markings on the planet are the belts of Jupiter. Along the Jovian equator is a light-colored belt, varying in width from 5,000 to 10,000 mi. On either side are two belts of reddish-brown color, followed by several similar, but narrower and less conspicuous, belts of red, brown, yellow, and tan tints in the higher latitudes. These belts vary in width and intensity, changing from day to day with occasional "storms" that produce marked alterations in a few hours.

In 1878 a great red spot appeared in the southern principal dark belt and was very conspicuous until 1881, being 30,000 mi. long and 7,000 mi. wide. First pinkish in color, it became a decided red, eventually fading and becoming less elongated. It has disappeared and reappeared; but, when absent, it leaves a hollow in the southern belt. Many other fainter spots have been observed, which change their structure much more rapidly. We can scarcely attribute these changes to seasons, for they bear no relation to the Jovian seasons. Furthermore, Jupiter's equator is inclined only 3°7' to its orbit; and, since the eccentricity is also small, the planet can have no decided seasons. Measurements indicate that the temperature of the surface is about 200° below zero on the Fahrenheit scale, which means that the clouds above Jupiter are not clouds of water vapor but are composed of some substance yet to be determined, of a much lower temperature.

SATURN

BEYOND the orbit of Jupiter is the most remote planet known to the ancients. Named after the god of time, sign ♄ (an ancient scythe), Saturn moves so slowly that it takes a generation—more precisely, 29.46 yr.—to trace its motion among the zodiacal constellations. As seen from the sun, every year and 13 days the earth passes Saturn and the latter is at opposition.

Through a telescope, Saturn with its wonderful rings seems the most interesting of all the planets. Without its rings, Saturn would resemble Jupiter, which is about 16 per cent greater in diameter. On both of these giant planets there are dark and light belts and spots of varying size, intensity, and duration. Near its equator, Saturn is yellowish in color, while the polar regions have a greenish tinge. The Saturnian day is only about half an hour longer than the Jovian, though, as was the case with Jupiter, the period of rotation increases with the latitude. Saturn's equator, however, is inclined almost 27°; and there are seasons on this planet—not, of course, like those of the earth, for the mean temperature of the planet is 360°, on the Fahrenheit scale, below that of the earth and, while our seasons are a mere 3 months long, Saturn's endure for $7\frac{1}{2}$ yr.! For more than 1,200 Saturnian days, the sun shines on each of Saturn's poles. Of all the planets for which such information is available, Saturn has the greatest oblateness, its polar diameter being more than 8,000 mi. shorter than its equatorial. Another distinction is that it has the smallest average density—only seven-tenths that of water; our model of Saturn in St. Paul, Minnesota, if constructed with the same average density, would float. Nine satellites have been observed to revolve about the planet—eight in an easterly direction and one, the most remote, westerly.

IN 1610, when he looked at Saturn with his small telescope, Galileo noticed that there was something peculiar about its shape; and he remarked, "Saturn is threefold." It seemed as though Saturn had two wings or appendages on either side to help it on its way. Then, as Galileo continued to watch the planet, the appendages disappeared. Forty-five years later, C. Huyghens established their true nature and, in an anagram which he later translated, stated his conclusions,

PLATE 10 SATURN, PHOTOGRAPHED BY BARNARD WITH 60-INCH REFLECTOR OF MOUNT WILSON OBSERVATORY

as follows: "It is surrounded by a ring, thin, flat, nowhere touching, inclined to the ecliptic." Twenty years after that, in 1675, Cassini found that the ring was divided into two by a gap, now known as *Cassini's division*. A faint inner ring called the *crêpe* ring was discovered two centuries later.

Modern measurements show that the rings are in the plane of Saturn's equator and are inclined about 27° or 28° to the ecliptic, the outer limit of the ring system being 86,300 mi. from the planet's center. Cassini's division terminates the outer ring, which is 11,100 mi. wide, the gap itself being only 2,200 mi. wide. Passing Cassini's division, we come to the brightest ring, 18,000 mi. in width, which merges with the inner edge of the *crêpe* ring, the ring system terminating at a distance of 6,000 mi. from the surface of the planet.

Shining entirely by reflected sunlight, as does Saturn and all the planets, these rings, with the shadow of Saturn clearly marked on them, have an appearance of solidity that they cannot possibly possess. Laplace pointed out toward the end of the nineteenth century that no solid ring whirling about Saturn could withstand the terrific gravitational forces to which it would be subjected, for, according to the Law of Gravitation, the force on the inner portion of the ring would be much greater than that on the outer edge and, consequently, a solid ring would be rent asunder; nor can the ring be fluid (as proved by Maxwell in 1857) but must be made up of disconnected particles, perhaps dustlike in character. That this is the case has been borne out in recent determinations of velocities for different portions of the rings. The rings do not rotate as a whole, but the particles in the outer edge move with a slower angular velocity than those in the inner; the former revolve about the planet in 13.7 hr., the latter in 5 hr., precisely the theoretical periods in which a particle would travel about the planet if it moved in a circle at the same distances. Furthermore, these rings are transparent; the planet is readily seen through the *crêpe* ring and, to a lesser extent, through the outer one, while stars are visible through the bright ring.

No noticeable gravitational influence of the rings on the nine satellites that lie beyond has been detected, so that the combined mass of the rings is probably at most a thousandth of that of the earth. Cassini's division is accounted for as an excellent example of celestial resonance. Have you ever noticed how, when a certain note of a piano or any other musical instrument is sounded, there sometimes is a sympathetic vibration from a vase or some such ornament? This is the phenomenon of resonance. Physicists tell us that such sympathetic vibrations will take place when the vibration rate of the ornament is either equal to, or a simple multiple of, that of the note. If a particle moved in Cassini's division, it would make twice as many revolutions as Mimas (period 22^h37^m), the closest of the known Saturnian satellites, and three times as many as Enceladus (period 32^h53^m); and, just as the vase vibrated sympathetically to the musical note, so the particle would be shaken out of Cassini's division by the gravitational action of Mimas and Enceladus.

VIEWING Saturn from the earth is practically the same as viewing it from the sun. At the Saturnian solstices the poles of Saturn are tilted toward the sun, and hence also toward the earth; and consequently, since the rings are in the plane of Saturn's equator, we see them at their maximum inclination, about 28°. But at the Saturnian equinoxes, the equator of Saturn passes through the sun and we see the rings edgewise. Since a year on Saturn is about 30 of our years the interval from vernal to autumnal equinox for this planet is 15 of our years; and thus every 15 yr. the rings are seen edgewise, viz., 1907, 1922, 1937, 1952, etc., and in the mid-intervals the rings are seen at maximum inclination. When the earth is exactly in the plane of the rings, they are barely seen in our most powerful telescopes, indicating that they are extremely thin, probably less than 20 mi.

We naturally wonder how Saturn obtained such rings. One suggestion is that they are the remains of a satellite that came too close to the planet, for, if such an object ventured into the region of the rings, it would be torn apart by the gravitational forces which would act upon it. Suppose such an annihilation happened in the past; the resultant particles would move at first in different planes, but repeated collisions would gradually force these objects to move practically in one plane. If the particles continued to collide, as would seem quite probable, then it can be shown that they would slowly approach the planet and eventually fall into its equator. When this happened, since the inner particles revolve faster than the planet rotates, a particle of the ring falling into Saturn's equator would accelerate the latter. At this point, it might be well to recall that the equator of both Saturn and Jupiter rotates faster than the middle latitudes. Perhaps this is the explanation.

But no matter what their origin might be, what a wonderful sight the rings would present if they could be viewed from the planet, spanning the sky in an immense arch and shedding a soft radiance over the surface of Saturn.

URANUS

IN THE year 1781, Sir William Herschel was examining some stars in Gemini when a small star, which seemed to be a definite disk, attracted his attention. He watched it a night or two, observing that it moved among the stars, and little realizing the importance of the object, he announced the discovery of a new comet; a few months of examination, however, plus computations of its orbit, revealed his error. The object was a planet—the first to be discovered since ancient times—and, since its distance from the sun was 19 astronomical units, this new object doubled the extent of the known solar system.

Herschel proposed to call it Georgium Sidus ("The Georgian Star") in honor of his patron, George III; but this found little favor. Various other names were suggested, his own included, before a decision was reached to call it Uranus,

THE SOLAR SYSTEM

after the god of the skies, which explains the significance of its sign, ⛢, an arrow pointing upward to the heavens.

On tracing its motion backward in time, it was learned that the planet had been previously observed but had been regarded as a star. Lemonnier at Paris had noticed Uranus on twelve successive nights; and, had he been a little more careful in recording his observations, he would doubtless have detected its planetary character. There is a story that one of Lemonnier's observations of Uranus

PLATE 11   URANUS AND FIVE OF ITS SATELLITES

was written on a paper bag, which had contained hair-powder purchased at a perfumer's! These earlier observations were of great importance in obtaining an accurate determination of the path of Uranus and, as we shall see, led to the discovery of Neptune.

TO THE naked eye, Uranus is barely visible on a clear moonless night, appearing as a star of sixth magnitude. It moves very slowly among the zodiacal stars, completing its circuit about the sun in 84 yr.; and it is no wonder that the ancients failed to notice it among the thousands of stars which are brighter than this planet. Through a telescope the planet presents a tiny disk decidedly green in color. Vague belts have been discovered; and therefore it is believed that this planet resembles Jupiter in physical structure, though much smaller in size, for its diameter is about 32,000 mi.

Rotation has not been directly observed by motion of spots; but other types of investigation indicate a rapid rotation, with a day only 11 hr. long, which is consistent with measurements of the planet's oblateness (the polar diameter is almost 3,000 mi. shorter than the equatorial). All measurements imply that the axis of rotation is approximately in the plane of its orbit, which, in turn, almost coincides with our ecliptic. Consequently, Uranus' equator lacks 8° of being at right angles to the ecliptic; and, in this plane and traveling in the same direction as the planet rotates, are its four known satellites: Ariel, Umbriel, Titania, and Oberon. Each is considerably smaller than our moon; and they journey about Uranus at distances varying from 120,000 to 360,000 mi., with periods of $2\frac{1}{2}$ 13 hr.

FREQUENTLY the statement is made that Uranus rotates, and its satellites revolve, westward. Though correct, this description is a trifle misleading. Instead, let us picture conditions on the earth, if its axis were parallel to that of Uranus; to do this, we increase the angle between the earth's equator and ecliptic by tilting its axis. Finally, the inclination is 90° and the earth's north pole is on the ecliptic; but let us not stop here—let us continue tilting in the same direction until our present north pole is south of the ecliptic and the earth's axis is inclined 8° to it.

If we then called our present north pole the "south pole," and the hemisphere containing the United States the "southern hemisphere," then east would become west and the sun would set in the east. But if we still called our present north pole the "north pole," then the sun would rise in the east and set in the west, just as it does now. At the time of the equinoxes the motion of the sun would be along the celestial equator; but at the solstices the sun would be almost overhead at one of the poles, describing a small circle only 8° in radius about the north celestial pole.

Under such conditions, the "land of the midnight sun" in the northern hemisphere would include all of North America and tourists would watch the sun reaching its lowest point due north on the horizon, as they passed through the Panama Canal. Increase our year more than eighty-fold—cut the day to less than half—diminish the sun in size until it is a mere point of dazzling light, equivalent in intensity to more than three thousand moons—and we have a better idea of conditions on Uranus.

NEPTUNE

FOR half a century after the discovery of Uranus, its path was carefully followed. How planets ought to move under the attraction of the sun was first determined, and then corrections were made for the influence of the other planets. In 1820, however, slight departures were noticed in the actual orbit of Uranus

THE SOLAR SYSTEM

and the theoretical one. As the years passed, these departures became more serious, and the Law of Gravitation and the laws of motion were at stake; many astronomers were wondering whether Newtonian mechanics could still be depended upon to predict the position of celestial objects. By 1840 the discrepancy was so great that, if a point of light had been placed at the predicted position of Uranus, this point of light and the planet itself would almost have appeared as separate points to the unaided eye. It had been suggested in 1820 that these unaccountable departures were due to the attraction of a more remote planet, and the problem was: "Given the departures, find the planet."

Two men attacked the problem independently of each other: J. C. Adams, of Cambridge, England; and V. J. Leverrier, of Paris, France. Adams transmitted his results to Sir George Airy, the then Astronomer Royal of England, who, unfortunately, had previously concluded that the departures alone would not determine the planet's distance from the sun, which explains his lack of enthusiasm for Adams' work. (Adams, as well as Leverrier, had assumed a distance about twice that of Uranus.) Leverrier was more fortunate, for he communicated his results on September 23, 1846, to J. G. Galle. This young German astronomer waited impatiently for a clear night; and finally, when he turned his telescope in the direction indicated by Leverrier—essentially the same position as that obtained independently by Adams—he found Neptune in less than half an hour—and only half a degree away from the predicted position! The dramatic discovery of the new planet was a vindication of Newtonian mechanics and one of its greatest triumphs.

SLOWLY, indeed, does Neptune move, taking about 165 yr. to complete its journey about the sun. Named after the god of the sea, its sign is Ψ, Neptune's trident. Its distance from the sun is only 50 per cent greater than that of Uranus—not twice that of Uranus, as Leverrier and Adams had guessed. (It can be shown, however, that this apparently poor guess would not greatly affect the prediction of its position in the sky.)

To us, Neptune appears a star of about magnitude 8, visible only through a telescope; and even with a powerful one, we have not, as yet, determined its rotation. What little information we have obtained indicates that it is similar in size and surface structure to Uranus. Only one satellite has been found, probably comparable in size to our moon; and, although its distance from Neptune's center is about the same as that between earth and moon, because of Neptune's greater mass its satellite revolves about it in a period of $5^d 21^h$.

PLUTO AND THE PLANETOIDS

AFTER the discovery of Neptune, many astronomers considered the possibility of detecting and locating more remote planets, if such there were, by their gravitational influence. If the separation of planets increases with distance from the sun, then the perturbations on the known planets due to more remote ones, unless these were of enormous mass, would necessarily be small, and this would also be true of their brilliancy. If Neptune's distance were doubled, its brightness to earthlings would be cut to approximately one-sixteenth its present intensity, because of the increase in the distance of the planet from the earth as well as from the sun. From studies of the motions of comets that recede far be-

PLATE 12 Neptune and Its Two Satellites. (McDonald Photograph.)

yond the orbit of Neptune, astronomers decided that trans-Neptunian planets actually existed.

Percival Lowell made a careful study of the motions of Uranus and Neptune and found that there were slight discrepancies in the predicted and observed positions. He attributed these to the action of a remote planet rather than to errors in observations, and from them he determined possible positions for this object. Part of the original program of the Lowell Observatory at Flagstaff, Arizona, was to find this hypothetical object; and 14 yr. after the death of Lowell, the diligent search of the Lowell observers was rewarded, when, on January 21, 1930, C. W. Tombaugh found the image of a new planet on a photographic plate. A careful study of Pluto's orbit, made possible by the discovery of its image on

THE SOLAR SYSTEM

plates taken prior to its recognition as a planet, revealed that its distance is considerably less than that predicted by Lowell. Its faintness (Pluto is of the fifteenth magnitude) indicates that it is probably of much smaller diameter than the earth, and this conclusion has been confirmed by recent measurements. Unless its density is unreasonably great, it could not have produced observable departures in the motion of Uranus; moreover, there is considerable doubt as to the existence of the discrepancies employed by Lowell, and we are compelled to conclude that the discovery of Pluto was due to careful observations rather than to a mathematical prediction. How often in the history of astronomy, as well as all other sciences, has man made a marvelous discovery when seeking something else!

PLATE 13 Two Photographs of Transneptunian Planet Pluto, Taken with 2-Foot Reflector, Showing the Change of Position among the Stars in 1 Day. Taken March 10 and 11, 1934.

The orbit of Pluto differs radically from those of the giant planets; whereas they move essentially in circles and identical planes, Pluto's eccentricity (0.25) and inclination to the ecliptic (17°) are even greater than those for Mercury. At perihelion, Pluto is closer to the sun than Neptune, though the inclination of Pluto's orbit is high and their paths do not cross; while at aphelion, it recedes to a distance of about 50 astronomical units. The shape and orientation of its orbit, plus its small size when contrasted with the giant planets, suggest that it might belong to another class of objects in the solar system—the planetoids.

THE discovery of the first planetoid, or asteroid, was the result of a mistake. In 1772, a German astronomer, J. E. Bode, noticed that the approximate distances in astronomical units of the then known planets could be obtained as follows: To a series of 4's add 0, 3, 6, 12, and so forth, each successive number after the second being twice the previous, and divide the result by 10. Thus we have

	4	4	4	4	4	4	4	4	4
	0	3	6	12	24	48	96	192	384
	0.4	0.7	1.0	1.6	2.8	5.2	10.0	19.6	38.8

On comparing these figures with the actual distances in astronomical units (Mercury, 0.39; Venus, 0.72; Earth, 1.00; Mars, 1.52; Jupiter, 5.20; Saturn, 9.54), it was conceded that Bode's numbers were fair approximations; but the number 2.8 apparently demanded explanation. When Uranus was discovered about 10 yr. later with a mean distance of 19.19 astronomical units, it made astronomers of the time consider this law seriously. If Neptune, with a mean distance of 30 astronomical units, had not been discovered (both Leverrier and Adams assumed Bode's number 38.8 in their calculations), the discovery of Pluto (mean distance, 39.6 astronomical units) would again have brought the law into prominence.

Just prior to the nineteenth century we find a society of twenty-four astronomers seeking a planet between the orbits of Mars and Jupiter. Curiously, it was a non-member, called G. Piazzi, who noticed on January 1, 1801, a faint object where none had been seen a few days earlier. Nor was it Bode's law that led to the discovery: the direct cause was an error of the press in putting an extra star in a catalogue; and when Piazzi was looking for this "printer's" star, he found Ceres! A few nights of observation showed that it moved among the stars and was therefore a member of the solar system. Illness prevented Piazzi from continuing his watch, and the news of its discovery traveled so slowly that the object became lost. Fortunately, the brilliant mathematician, C. F. Gauss, took up the problem of locating it and developed a method, still used today in a modified form, of determining the future position of a comet or planet, as well as the characteristics of its orbit, given only three observations of its apparent position. Gauss's calculations led to the rediscovery of Ceres. Its mean distance of 2.77 astronomical units seemed to imply that it was the long-sought-for planet, though somewhat undersized, for present measurements give Ceres a diameter of 480 mi.

A year later another dwarf planet was found—Pallas (diameter 300 mi.); in 1804 Juno (diameter 120 mi.) was discovered; and in 1807 Vesta (diameter 118 mi.), the only known planetoid that is ever visible to the eye. No more planetoids were added to the list until 1847; but thereafter at least one a year has been found, and now they number more than a thousand. With the development of photography the rate of their discovery has increased enormously. The motion of the planetoids among the stars is in general so rapid that a "photographic trap" can be set for them. For instance, in photographing a group of people, if one person moves during the exposure it is a simple task to locate the guilty party after the plate is developed; so if the "fixed stars" are photographed, a wandering planetoid leaves a tiny streak on the plate.

THE SOLAR SYSTEM

A PRINTER'S error, therefore, began an enormous program of observations and calculations which has greatly embarrassed the astronomer. The naming of these tiny objects, some measuring only a fraction of a mile, at first caused arguments comparable with those that occur at the birth of a child. Names were chosen from Greek mythology, the masculine ones of course being reserved for the more exceptional planetoids, but soon even the supply of feminine names was exhausted. We therefore find planetoids rejoicing in the names of Shakespearean and Wagnerian heroes, as well as being called after favorite desserts; one astronomer is said to have financed an eclipse expedition by selling his rights of naming a planetoid. Nowadays, however, these objects are better identified by a number indicating the order of their discovery.

Because of the tremendous labor involved, the orbits of only about two-thirds have been accurately determined; and there is no doubt that many planetoids have been rediscovered and renamed. Just as the planets travel eastward about the sun, so do the planetoids select this same direction. Their orbits, however, vary greatly both in eccentricity and inclination—some are circular, others have an aphelion distance four or five times their perihelion; inclinations sometimes exceed 45°. Most of the now known planetoids lie between the orbits of Mars and Jupiter, the average distance in astronomical units of the entire group from the sun being about 2.8, which was the number in Bode's series. Nevertheless, we must be careful not to draw too hasty conclusions from our present knowledge, for it is possible that there are great numbers beyond the one planetoid discovered which recedes to a heliocentric distance of about 10 astronomical units. Many astronomers believe that these objects extend out to the limits of the solar system and that Pluto, because of its highly eccentric and inclined orbit, is really a giant planetoid. When we consider that there are nearly fifteen million stars brighter than Pluto, it seems not unlikely that many such objects could exist at these distances and escape detection.

ONE reward for the enormous amount of work caused by the planetoids was the discovery of Eros in 1898. At its closest it comes within about 13,000,000 mi. of the earth, though it recedes to a distance of around 24,000,000 mi. beyond the orbit of Mars; and its importance lies in the fact that it provides us with a means of determining the dimensions of the solar system. Though Kepler was able to draw the plan of the then known solar system, just as an architect draws the plan of a house, he was ignorant of the scale. Given the actual magnitude of only one distance, the scale is determined. Aberration (see chap. 1) enables us to find one such distance: the mean distance of the earth and the sun. The geometrically determined minimum distance of Eros provides us with another independent, and more accurate, method of obtaining the magnitude of the astronomical unit and the scale of the solar system.

PLATE 14 TRAIL OF EROS, FEBRUARY 16, 1931; EXPOSURE, 1 HR. 30 MIN.

THE SOLAR SYSTEM

Two planetoids of great interest were discovered in 1932. Amor, a small faint object, probably only 1 or 2 mi. in diameter, was just 10,000,000 mi. from the earth on March 22, 1932. The other object, as yet not named, but designated "1932 H.A.," is of about the same size and was discovered by Karl Reinmuth on April 27, 1932. In May, 1932, its distance from the earth was about 6,500,000 mi., and no closer approach of a planetoid has ever been observed. Furthermore, it has been calculated that it is possible for this object to come within 3,000,000 mi. of us. Heretofore all planetoids have had orbits outside that of the earth, but Reinmuth's object actually *loops* the orbits of Mars, Venus, and the earth!

COMETS AND METEORS

WE COME now to the apparently unruly members of the solar family—the COMETS. The suddenness with which they appear in the sky, their terrifying size with tails stretched out almost across the heavens, and the weird forms they assume have made them a source of fear and superstition. No wonder man in the past regarded these "hairy stars" as—

> Threatening the world with famine, plague and war;
> To princes, death; to Kingdoms, many curses;
> To all estates, inevitable losses;
> To herdsmen, rot; to plowmen, hapless seasons;
> To sailors, storms; to cities, civil treasons.

Their large glowing heads often exceed the sun in diameter, the average-sized being comparable to Jupiter, while their streaming tails (comae) can sometimes be traced for more than a 100,000,000 mi. In the midst of the nebulous head there usually is a nucleus (or nuclei), a brilliant starlike point to the eye, though actually a few hundred miles in extent. Comets are not confined, as are the planets, to the zodiacal constellations but may appear anywhere in the sky and, excluding those few whose paths are accurately known, unannounced.

Kepler remarked that the comets were as numerous as the fish in the sea; and, though but a thousand comets have so far been observed and recorded, some astronomers believe they number in the hundreds of thousands. Every year from three to eleven are seen, of which approximately one-third are definitely identified as returns of previously discovered comets; in fact, it is the belief of most astronomers that the newly discovered comets have also made past visitations to the sun. Unfortunately, a comet is telescopically visible only when within a few astronomical units of the sun, and from a small observed portion of its path an attempt is made to reconstruct the entire orbit. Where reliable observations are available, it has been established that comets travel in very elongated ellipses (neglecting the perturbative effects of the planets) in accordance with Kepler's laws, and there seems no conclusive evidence of any comet having entered the solar system from outside. The computed periods for these orbits range from a little more than three years to a few thousand years, and their inclinations have all values from 0° to 90°. With a few exceptions, of which Halley's comet is the most notable, all comets move about the sun in the same direction as the planets.

ABOUT thirty comets have been discovered which recede to the orbit of Jupiter—Jupiter's "family" of comets. They do not belong to Jupiter in the same sense as do its satellites, for one of the foci of their elliptical paths is the sun and it is only their *aphelia* which lie close to Jupiter's orbit. Just two comets have been found with aphelia near the orbits of Saturn and Uranus. Neptune has a family of seven comets, of which the famous Halley's comet is a member. The remaining comets recede to even greater distances, as we follow their courses mathematically; perhaps they, too, are members of cometary families of still undiscovered planets.

When first seen, a comet appears as a faint spot of light upon the dark background of the sky, gradually increasing in brightness as the object approaches the sun. Occasionally—on the average, two or three times in a century—we see a comet that is visible even in the daytime. At a distance of a few astronomical units from the sun, a tail begins to form on a large comet, such as Halley's, streaming outward from its head and directed away from the sun. This tail grows in splendor as the comet nears perihelion; additional tails often shoot out from the head and afterward disappear. Obeying the Law of Areas in its motion, a comet sometimes whips around the sun in a few hours at a distance of only a million miles. Frequently it is torn apart by the tidal action of the sun: the great comet of 1882, after its close perihelion passage, was attended by four smaller cometary bodies, and its nucleus was divided into five. This comet itself pursued a path very similar to that of the great comets of 1843, 1880, and 1887; and it is suspected that all four were originally one great comet which was broken up at some former solar approach.

As the comet recedes from the sun, its tail goes ahead, being always more or less directed away from the sun. All evidence indicates that the tail is formed at the expense of the comet by some repulsive solar action—light pressure or electrical repulsion. Its brightness lasting from a few days to about 2 months, it loses whatever tails it had and gradually fades away and disappears as its heliocentric distance slowly increases.

WITHOUT doubt the earth has passed many times through the tails of giant comets with perihelia inside the earth's orbit, as was the case with the famous Halley's comet on May 18, 1910. The visitations of this celebrated member of Neptune's family, in intervals of about 75 yr., have been found dating back to 240 B.C., though Halley, a friend of Newton, at its appearance in 1682, was the first to predict its return. It was this comet that was looked upon with dread as

PLATE 15 COMET 1910 II, HALLEY, MAY 4, 1910

the forerunner of the victory of William the Conqueror in 1066; its appearance in 1456 followed the fall of Constantinople (1453), and the prayer of the Christian world was: "Lord, save us from the Devil, the Turk, and the Comet." Nevertheless, our encounter with its tail in 1910 produced no observable effects. (We admit the World War began in 1914, but historians now trace such events to the natural bellicosity of the human race and not to celestial phenomena.) Nor was any catastrophe anticipated, for, while the volumes of comets are enormous, their masses are insignificant in comparison with the planets. The evidence for this conclusion is that, on the occasion of some close approach of a planet and a comet, the orbit of the latter is materially changed; but no observable alterations have been produced in the motion of the planet. More than this, when the great comet of 1882 and Halley's comet transited the disk of the sun, they were completely invisible. As for a comet's tail, its density is probably far below the best vacuum attained in any physics laboratory, and therefore we need not fear the deadly carbon monoxide and cyanogen detected in the analysis of the light from comets. Their source of illumination is the radiant energy with which they are impregnated by the sun. Part of this is reflected, part is absorbed to be radiated once more, so that a comet—and this is particularly true of its tail—resembles a luminous gas of nitrogen and various compounds of carbon. When near perihelion, there also appears light identified as that emitted by sodium vapor.

THE possibility of a collision with the head of a comet no longer alarms us. In June, 1921, the earth narrowly escaped collision with the Pons-Winnecke comet; but, while we have no record of an impact with the head of a comet, it is believed that we are repeatedly encountering the remains of former comets.

In 1826 Biela's comet was discovered and was found to have the short period of 6.6 yr. It was observed again in 1832, missed in 1839 because of its unfavorable position, and observed once more in 1846. That year the nucleus divided into two parts, which gradually separated; and in 1852 the twin comets reappeared at a considerable distance apart. Biela's comet has not been seen since 1852; but in the year 1872, as the earth crossed the track of the lost comet, there was a magnificent display of "shooting stars" or meteors. This phenomenon has been repeated at similar passages of the earth across the orbit of this particular comet.

METEORS are so common an occurrence that a description of their appearance in the sky is almost unnecessary. Though most of them equal the brightness of the stars visible to the eye, a few equaling or exceeding the brilliancy of Venus, estimates of their mass (based on velocity and the radiation liberated) indicate that hundreds, if not thousands, of these "shooting stars" could be held in one's hand. Their height is found by methods analogous to that employed in determining the distance to the moon; and they appear, on the average, 80 mi.

THE SOLAR SYSTEM

above the earth's surface, to disappear at altitudes of from 30 to 60 mi. The luminosity of a meteor is due to heat produced by friction in passing with a high velocity through the rarefied upper atmosphere.

A single observer may count anywhere from six to a hundred meteors in an hour; but, inasmuch as he is able to see only those meteors within a few hundred miles of his position, the total number striking the entire earth must be many times greater. It is estimated that the daily total cannot be fewer than several millions. At times this number is greatly augmented, and we have meteoric showers similar to the display at the encounter with the remains of Biela's comet.

AN EXCEPTIONAL—and we might say, terrifying—shower of meteors was seen on this continent during the small hours of November 13, 1833. Single reliable witnesses estimated that they saw thousands of "shooting stars" hourly, some of which left trails that persisted for minutes. Careful observers noted that, although metors fell in all portions of the sky, when their trails were prolonged backward they all intersected at the same point—the *radiant* as it is called. Similar radiant points have been located during other meteoric showers, and the explanation is that the "shooting stars" are traveling essentially in parallel lines. (That parallel lines appear to intersect when extended indefinitely is obvious, when we look at a large number of long parallel railroad tracks.) The position of the radiant gives us the direction in which the meteors of a shower are moving.

In the years following the great showers of 1833, it was noted that relatively smaller showers occurred on or about the same date with the same radiant point, namely, in the constellation of Leo, from which they derive their name—the Leonids. Then it was recalled that on November 11–12, 1799, the naturalist Alexander von Humboldt had seen a shower similar to that of 1833 and, consequently, a repetition of these terrific showers was predicted for 1866. The prediction was fulfilled, but the meteors were not quite so numerous as in 1799 and 1833. Though a considerable number fell in the beginning of this century and in 1932 and 1933, the displays of 1799 and 1833 have not as yet been equalled.

INASMUCH as the Bielids, the showers associated with Biela's comet, are similar to the Leonids, it is believed that the latter are also due to some comet, the orbit of which we cross on November 12–14. Probably the exceptional showers that occur in cycles of $33\frac{1}{3}$ yr. (the Leonids have now been traced back to 902 A.D.) are encounters with a large fragment of a comet, which fragment is traveling in an ellipse about the sun in a period of $33\frac{1}{3}$ yr., while the lesser Leonid showers are due to cometary remains strewn over the comet's entire orbit. Moreover, Tuttle's comet of 1862 and Tempel's comet of 1866 pursue paths similar to that computed for the meteors of the Leonids; in fact, all the meteoric showers listed in the accompanying table have been associated in this manner with some comet or comets.

Radiant in Constellation	Meteoric Shower	Date of Shower
Lyra	Lyrids	Apr. 20–21
Perseus	Perseids	Aug. 10–11
Draco	Giacobinids	Oct. 9
Leo	Leonids	Nov. 14–15
Andromeda	Bielids	Nov. 23

While the Leonids and Bielids do not occur every year, the Lyrids and Perseids have about the same yearly intensity. The best time of night to observe these showers is when their radiant point is high in the sky.

ALL meteors, however, do not belong to these definite swarms; there is a vast number of sporadic meteors, i.e., meteors without radiants. Recent measurements indicate a sharp demarcation between the velocities of sporadic meteors and those belonging to showers. Relative to the sun, the latter have velocities of less than 27 mi. per sec., while the sporadic meteors have velocities greater than this, many traveling even 80 mi. per sec. As 27 mi. per sec. is the very maximum speed for a permanent member of the solar system at the earth's distance from the sun, this and other evidence indicates that these sporadic meteors have had a cosmic origin. Whether or not the comets themselves were originally visitors from outside the system and were captured by the combined gravity of a planet and the sun is still an open question.

Now and then we observe exceptionally bright meteors, rivaling Venus and even the moon, which from their appearance are called FIREBALLS. They are often dissipated in a succession of explosions followed by loud reports and, owing to air resistance, their paths are generally erratic and not in straight lines.

WHEN such celestial visitors chance to reach the earth's crust, they are known as METEORITES. Probably the only distinction between meteors, fireballs, and meteorites is one of size; and the meteorites we examine in museums may very possibly be visitors from beyond the solar system. Most of those that are seen to fall (the aërolites), when analyzed, are found to consist of masses of stone (i.e., limestone, magnesia, or silicon) mixed with ferric granules. A small percentage (the all-metal, or siderites) are of nearly pure iron alloyed with nickel. Slight traces of copper, phosphorus, cobalt, and sulphur are also found, as well as quantities of occluded gas—carbon dioxide, carbon monoxide, hydrogen, and nitrogen. Though no new elements have been discovered in them, their peculiar structure and the presence of compounds not found in rocks indigenous to the earth identifies them. Most of the meteorites in museums are metallic ones which fell in the distant past, and their particularly hard structure has withstood the erosion which disintegrated the stony ones.

The number of instances in which meteorites have been seen to fall and later recovered now averages about four per year, the total number striking the earth annually being probably several hundred. Nevertheless, there have been a few mortalities attributed to such objects. Chinese records mention one that fell

THE SOLAR SYSTEM

about 2,500 yr. ago and killed ten men. These celestial visitors were held in great awe by the ancients: in Rome a meteorite was worshiped as "the Mother of the Gods"; another one was excavated in a ruined Aztec temple.

The mass that falls is sometimes a single piece, but more often it is composed of many fragments. At the Pultusk fall of 1869 the number of meteorites was estimated to exceed one hundred thousand, though most of them were mere pebbles. On the other hand, the American Museum of Natural History in New York possesses an iron meteorite weighing about 36 tons. It is the largest single meteorite discovered to date and was brought back by Admiral Peary from Greenland. How large are the metallic meteorites buried in Meteor Crater in Winslow, Arizona (see chap. 4), has not as yet been determined. We also wait for more information concerning the devastating swarm of meteorites that fell on June 30, 1908, in North-Central Siberia.

THE ZODIACAL LIGHT

SOON after dusk on any clear moonless night—but preferably in March or April for the middle-northern latitudes—we see a vague triangular patch of light. Its base is formed by the western horizon, where it is wider and brighter than the Milky Way; but it gradually tapers to about 2° as we trace it a third of the way across the sky. Occasionally, keen eyes may follow a very faint thin band to the eastern horizon. As the ecliptic passes down its center, we call this the ZODIACAL LIGHT. The zodiac is where we find the planets, and the zodiacal light is accounted for as being the reflected light of countless small particles circulating about the sun in orbits near the ecliptic.

At a point in the sky directly opposite the sun is a still fainter oval patch, a few degrees wide and about 10° long. This GEGENSCHEIN (counterglow) is claimed by F. R. Moulton to be due to a condensation of these particles into a dynamical whirlpool, caused by the combined attraction of the earth and sun.

THE SUN

DOMINATING the entire solar system is the Sun, to the human race the most vital celestial body. We acknowledge its importance in that, *imprimis*, solar radiation is our only source of energy worthy of mention; *secundus*, it is the dictator of the motion of the Earth and all the other planets, as well as of the planetoids and comets; *tertius*, the Sun is a star.

By comparisons with our own relatively insignificant planet, we strive to derive a picture of the Sun's immensity. Its apparent angular size and its distance from us imply a diameter 109 times that of the Earth and a volume of 1,300,000 times. The methods of celestial mechanics (chap. 5) assign to it a mass 332,000 times that of our planet, i.e., about 700 times the combined mass of all the planets; consequently, the average density of the Sun is about 0.25 that of the Earth. These and other statistics pertaining to the Sun are tabulated herewith for future reference. The derivation of certain of these quantities will be considered later. There is no need to speculate on the difficulties man would encounter were he transported to a body as large and massive as the sun; the force of attraction would multiply his earthly weight twenty-eight-fold, and his puny frame would collapse under this unwonted stress.

Mean diameter	864,000 mi., or 1,390,600 km.
Mass	1.32×10^{25} lb., or 5.97×10^{27} gm.
Mean density	1.41 (water = 1)
Surface gravity	28 times that of Earth
Velocity of escape	383 mi./sec., or 617 km./sec.
Period of rotation	25–34 days
Inclination of equator to ecliptic	7°
Equatorial velocity	2.5 mi./sec.
Energy radiated from surface	70,000 h-p. per sq. yd., or 63,000 kw. per sq. m.
Total energy radiated	9.07×10^{25} cal./sec.
Surface temperature	6,000° C.

Lest there be disastrous consequences, we repeat our warning not to attempt to look at the sun with an ordinary telescope; a piece of paper is quickly set on fire when placed near the eyepiece of a telescope directed toward that body. Although modern astronomical studies of the sun are made mostly from photographs on fine-grain plates, many of the solar phenomena can be observed by the amateur with very inexpensive equipment. By varying the position of the eyepiece of a small telescope (focusing), a sharp image of the sun may be formed and studied on a white cardboard screen.

THAT intensely bright circular disk that we see with or without optical aid is called the PHOTOSPHERE (light sphere), for it is from this surface that most of the solar energy is radiated. All of the visible sun, of course, is in a gaseous state; and, until we reflect upon the high value of the sun's surface gravity, it might seem strange that the edge of this "light disk" is so sharply defined. Moreover, at the sun's distance from earth, a gradual change of density through a depth of 50 mi. would pass unnoticed. We have no wish to imply that the photosphere is of uniform density. On the contrary, this surface appears to have a mottled or granulated structure, the granules averaging 500 mi. in diameter and being considerably brighter than the remainder of the photosphere. If two solar photographs are taken a fraction of a second apart, as a rule the granules on one cannot be identified on the other. They are regarded by some authorities as the tops of intensely hot metallic clouds, rising from the sun's interior.

Frequently, considerable agitation is seen in the photosphere; and relatively dark spots appear, covering, at times, thousands of millions of square miles. Occasionally a SUN-SPOT is so immense that it can be seen through a dark glass with the naked eye, or when the brilliant light of the sun is dimmed by a heavy mist.

A TYPICAL spot consists of a dark center (the UMBRA), surrounded by an irregular ring (the PENUMBRA) of a brightness intermediate between the umbra and the adjacent photosphere. On photographs the umbra has the appearance of a black chasm, but it is only dark by comparison; actually, if the spot could be removed from the sun, it would appear intensely bright. Very often an extremely brilliant bridge (or bridges) might extend across the entire spot, separating it into two parts. In the sun-spot neighborhood, the photospheric brightness is usually above the average; and close to the spots are bright patches called FACULAE, which, however, are also observed when no spots are visible. The umbra of a large spot may have a diameter of 50,000 mi. and the penumbra may extend another 100,000 mi.; the minimum size is unknown, for our telescopic studies are limited to those spots which are greater than 100 mi. in extent.

The student possessing even a small telescope will find the behavior of these spots extremely fascinating. They are relatively short-lived. Most of them appear and disappear in the course of a few days; others persist for a couple of months; but, in one recorded instance, a spot remained for 18 mo. The majority appear in groups —two large spots with a number of smaller ones in attendance, the latter being the first to vanish.

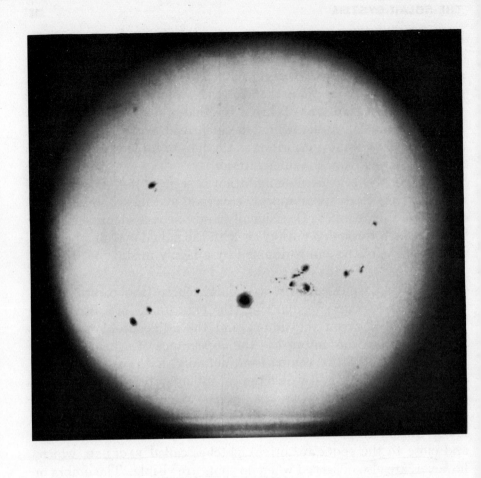

PLATE 16 SOLAR DISK WITH MANY SPOTS OF UNUSUAL SIZE, NOVEMBER 30, 1929. TAKEN WITH 40-INCH REFRACTOR

As the sun turns majestically on its axis, the spots are carried along across the apparent solar disk, enabling us to measure the period of rotation. We should, however, say *periods* of rotation, for measurements based, not only on sun-spots, but also on the apparent motion of bright clouds of calcium- and hydrogen-vapor (called FLOCCULI) and of the previously mentioned faculae, indicate that the angular rate of rotation decreases from equator to poles. (Incidentally, the spots themselves are never seen near the sun's polar regions and are always within 45° of the solar equator.) A point on the sun's equator completes its circuit eastward in 25 of our days, while the polar regions make one rotation in 34 days—a phenomenon similar to that observed in the planets Jupiter and Saturn. The plane of the solar equator is inclined about 7° to the ecliptic, and the sun's north celestial pole would be at a point midway between Polaris and Vega. The motion of these spots is not wholly one of rotation, for they drift relative to the photosphere and to each other. This drift may occur in any direction; but it is usually greatest in an easterly or westerly one, and sometimes amounts to several thousand miles in 24 hr.

All evidence indicates that a sun-spot is a funnel-shaped vortex or cyclone, within which gaseous matter streams upward and outward. As the gases advance, they expand rapidly, this sudden expansion producing, as is experimentally verifiable, a considerable cooling. The emitted gases flow more or less outward from the spot.

IT IS difficult to account for their periodicity. Since Galileo made the then blasphemous statement that the sun had spots, voluminous records have been collected concerning the number of, and the total area covered by, sun-spots. Though the daily number varies irregularly, the total number in a year oscillates in a well-defined manner. On the *average*, periods of maximum spottiness occur every 11.2 yr., though the interval from one maximum to the next has been observed to vary from 8 to 17 yr. During the first third of this century, the maxima and minima were found to be as follows:

Maximum.......	1905	1917	1928	
Minimum........	1901	1913	1923	1933

There is a remarkable connection between certain terrestrial phenomena and sun-spots. As a rule, the needle of a magnetic compass does not point due north; and, by comparing its direction with that of Polaris, we find on the eastern shores of the United States that the "north" defined by the compass is several degrees west of north, while on the Pacific Coast it is east of north. Moreover, the direction of the needle oscillates daily over a few minutes of arc. Now and then the "north" of the compass moves through an arc of several degrees within an hour, indicating considerable changes in the earth's magnetism. These MAGNETIC STORMS are most frequent at the time of sun-spot maxima, and occasionally they are so "violent" that induced electrical currents interfere with the operation of telegraph lines.

By "magnetic storms" we do not mean "thunder storms." Associated with magnetic storms are electrical displays of altogether different character from lightning, namely, the POLAR AURORA (i.e., the aurora borealis or northern lights) and AURORA AUSTRALIS. The gorgeous waving curtains and pulsating streams of soft greenish light, with their flashes of rose and violet tints, are due to discharges of electricity through the upper atmosphere. These "neon" signs of the sky are at altitudes many times that of the highest thunder cloud, and three or four times that of the highest meteoric flash; in fact, measured heights of aurorae inform us that our atmosphere extends at least 400 mi. from the earth's surface.

Not only are magnetic storms and aurorae particularly noticeable at sun-spot maxima, but—and this is important for broadcast enthusiasts—it is suspected that radio reception will not be as clear at such times.

WE SEEK the bridge that links the sun-spots with those terrestrial phenomena. Perhaps the connection is made through the PROMINENCES, those enormous masses of luminous gases which ascend to hundreds of thousands of miles and which are at times quite conspicuous to the unaided eye when the moon blots out the light of the solar disk at an eclipse (chap. 4). But the study of prominences is not altogether limited to the brief period of totality

THE SOLAR SYSTEM

at the time of a solar eclipse. Though the brilliant sky makes it quite invisible, the rose-colored light of a prominence is a composite of pure colors, chiefly due to incandescent hydrogen, helium, and calcium, which may be singled out from the light of the photosphere. Using an instrument based on the principles of the spectroscope, prominences have been photographed; and their behavior has been recently recorded on motion-picture film.*

They fall into two classes: QUIESCENT PROMINENCES, which are seen on the sun's limb in the form of massive pyramids or pillars, frequently saw-toothed in appearance; and ERUPTIVE PROMINENCES, which resemble rockets, jets of flames, and complicated arches. Whereas the quiescent type remain essentially unchanged for days and months, the eruptive prominences leave the sun's surface with velocities of 100–200 mi. per sec., and their forms are often completely altered in a few minutes. Though the quiescent prominences are rarely, if ever, associated with spots, the eruptive can usually be definitely connected with some particular spot; and, as might be anticipated, this spot is generally in its youth and more turbulent. Moreover, both types of

* See Bibliography.

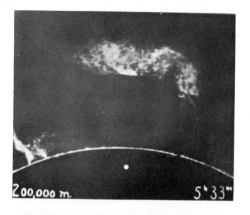

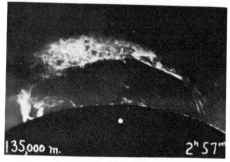

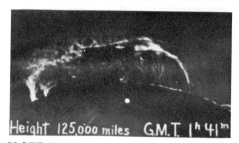

PLATE 17 Rise of Very High Prominence of May 29, 1919. Dot represents size of earth

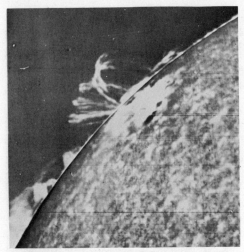

PLATE 18 COMBINATION OF PROMINENCE AND DISK

prominences are most numerous at sun-spot maxima; and it is suspected that not all the matter shot out during these violent storms returns to the sun but that, in the "explosion," some particles acquire velocities sufficient to escape the sun's gravity, i.e., greater than 383 mi. per sec., the sun's velocity of escape. After journeying a day or so, a very small percentage of these reach the earth and, as they enter our atmosphere, may produce a display of the aurora borealis. Perhaps it is the sun we must thank for supplying the electrons that bombard the atmospheric particles, to form that hypothetical layer of ionized air (the Heaviside layer), which is said to reflect the radio waves back to earth, permitting the inhabitants of this curved globe to "get distance."

THE prominences appear to rise from the CHROMOSPHERE (colorsphere), a layer of lighter gases, mostly hydrogen and helium, covering the sun to a depth of several thousand miles. This scarlet-colored gaseous envelope is separated fom the photosphere by the REVERSING LAYER, a layer of denser gases which gradually merges with the chromosphere at an altitude (measured from photosphere) of several hundred miles.

It is noticed in solar photographs that the interior of the disk appears brighter than the edge. The explanation of this is that the light we see from the center of the photosphere has the shortest path through the reversing layer and, consequently, less of it is lost on the way. This absorption of light is not the same for all colors, and each atom in the reversing layer makes its own selection. In the laboratory we duplicate this selective absorption of light and

THE SOLAR SYSTEM

find that atoms of the same element make their own particular co-optation, this seeming predilection for certain colors identifying the element. We are therefore able to discover just what elements on earth are also present in the reversing layer of the sun; and, of the ninety-odd elements recognized, upward of 60 per cent have been positively identified on the sun. The list includes hydrogen, oxygen, magnesium, iron, silicon, sodium, etc., and even such heavier elements as silver. It should be noted that this applies only to the composition of the sun's *atmosphere*, for we can say nothing about the elements in its interior; nor can we state definitely that any known element is *not* present in the sun. It is interesting to note that helium gas, now commonly used in dirigibles, was "discovered" on the sun in 1868 and identified on the earth more than 20 yr. later.

Above the chromosphere is the CORONA, extending outward at least 300,000 mi. from the sun's surface, though streamers may be followed for 200,000 mi. more. These measurements, as well as all observations of the corona, can be made only during a solar eclipse (chap. 4). The inner portions have a slightly yellowish tinge and the outer fringe is pearly white, while its total brightness is equivalent to one-half that of the full moon. Though its complex shape varies with the sun-spot cycle, its brightness is not materially affected.

FINALLY we consider what to us is probably the most important aspect of the sun—its heat. Though we are situated more than 90,000,000 mi. away from this source of energy, the power continuously supplied to the entire earth is: 126,000,000,000,000 h-p. In order to grasp the significance of this figure, we divide it by the population of the world and allot to each individual 60,000 h-p., i.e., one-fifth the installed capacity of Muscle Shoals.

Huge as these quantities are, they become absurdly small when compared with the solar radiation. Of the total energy emitted in all directions by the sun, our earth intercepts less than one part in two thousand millions! Expressed in calories (1 cal. equals the amount of heat required to raise 1 gm. of water 1° C.), the total solar radiation is 9.07×10^{25} cal. per sec.; i.e., 89,000 cal. per min.

PLATE 19 CORONA, JUNE 8, 1918, AT MATHESON, COLORADO

THE SOLAR SYSTEM

are liberated by each square centimeter of the sun, or 70,000 h-p. per sq. yd.

Knowing the rate of radiation, it is possible to estimate the temperature of the radiating surface. Various methods assign to the sun's surface a temperature of about 6,000° C. (11,000° F.). This temperature is so high that it would volatilize any known substance, and may be compared with that of the electric arc (3,700° C.) or the recently obtained high temperatures in electric furnaces (6,000° C.). The internal constitution of the sun can only be deduced from theoretical considerations. According to Sir Arthur Eddington, the temperature of the sun's center must be of the order of 10,000,000° C.

WE WONDER what *is* the source of all this energy. Inasmuch as the sun's mass is 1.985×10^{33} gm., we might hastily advance the hypothesis that it is simply a huge burning mass of carbon and oxygen—a ridiculous assumption, for such a combination would provide radiant solar energy for only another 1,500 yr. To obtain a better insight into what is demanded, let us divide the total number of calories radiated in a year by the total mass of the sun: the sun radiates 1.7 cal. per yr. per gm. Now, it is generally accepted that the earth is at least a billion years old, and during this period the sun has materially maintained this rate of radiation. Apparently, more than a billion calories have been generated per gram of matter!

Experiment after experiment has failed to disclose the sun's secret. A theoretical formula, which we owe to Einstein, makes energy and mass equivalent and states that a gram of matter, no matter what its constituents, corresponds to 2.15×10^{13} cal. The implication of this theory is that matter itself can be annihilated and an incomprehensible energy liberated; for example, the ash of a cigarette is equivalent to the heat required to melt many hundreds of tons of iron. If this hypothesis be correct, the sun is losing 4,000,000 tons of mass each second and, barring any outside source of matter to compensate, will decrease in mass one-thousandth of

1 per cent in 150,000,000 yr. However weird this theory sounds, it is at present the only adequate means we have to account for the sun's heat.

IN THE previous pages, we have given the salient features of what remained after the close approach of two stars. It was not our intention to show how the various peculiarities of the planets are accounted for in the different theories of the solar system. For this we refer the reader to the works of the originators of the planetesimal hypothesis—Chamberlin and Moulton—and the tidal theories of Jeans and Jeffreys.

CHAPTER 7

THE SIDEREAL UNIVERSE

Great fleas have little fleas upon their backs to bite 'em,
And little fleas have lesser fleas, and so *ad infinitum*.
And the great fleas themselves, in turn, have greater fleas to go on;
While these again have greater still, and greater still, and so on.
—DE MORGAN: *A Budget of Paradoxes*

THERE remains the task of describing modern conceptions of the universe. Twenty-five hundred years ago a simple answer satisfied man's questions concerning the Beyond. But no longer are we permitted to dispose of the stars by maintaining that they are holes in the dark firmament through which shines the pure light of paradise. Nor can we conscientiously terminate our discussion with the remark that the stars are other suns; three centuries of telescopic observations indicate that such a single classification is hopelessly inadequate. In a book such as this, we can present only a very abbreviated sketch of the nature of the world outside our solar system, as it is disclosed by modern astronomical instruments.

As seen from the earth, Sirius, the brightest star in the sky, is no brighter than a thousand candle-power light 6 mi. distant. The faintest star visible to the unaided eye, about magnitude 6, appears of the same intensity as a single candle 7 or 8 mi. away, while the faintest stars detectable by our largest telescopes, in conjunction with photographic plates, may be compared in brilliancy with single candles several thousand miles above the surface of the earth. It therefore seems incredible, with so little light available on the subject, that pages could be devoted to the physical aspects of a single star; and it is not strange that most of us are anxious for an opportunity to peer through a mammoth telescope.

NEVERTHELESS, one's first glimpse of a star through a telescope such as the 40-in. refractor of the Yerkes Observatory may be somewhat disappointing. Through this telescope—it has the world's largest objective lens (40 in.), which collects and concentrates the starlight on the eyepiece at the other end of the 60-ft. tube—the image of the star is still a mere point of light, though its brilliancy is increased thirty-five thousand times. This corresponds to a decrease of eleven or twelve magnitudes, so that a star barely visible to the naked eye is transformed into an object excelling the planets Venus and Jupiter in brightness. Many people, however, fail to appreciate this accomplishment, because the instrument does not reveal the star as a disk with definite markings.

One of our former statements requires modification. We remarked that the telescopic image of a star is a point of light; but, mathematically, a point possesses no dimensions. Stellar images, however, under ideal conditions, are circular disks of finite diameter. These disks are *spurious*, for it can be shown, by the theory of optics, that the image of a luminous point in an optically perfect telescope will not itself be a point but will be a *spot*, brightest at its center and fading away at the edges, surrounded by a series of faint concentric rings. To illustrate, let us consider two points of light: the image formed by the large objective lens consists of two such *diffraction patterns*, as they are called. If the two points are sufficiently close together, the corresponding spots will overlap and, given only the telescopic appearance, we might conclude that the two points were but one.

PLATE 20 THE 40-INCH REFRACTING TELESCOPE, WITH FLOOR AT LOWEST POSITION

IN THIS manner, we fail now and then to resolve telescopically a BINARY STAR, i.e., a system of two suns revolving about each other, into its two component sources of illumination. We therefore describe the *resolving power* of a telescope in terms of the minimum angular distance that two points of light can be apart and still be recognized as separate sources of illumination. This angle depends on the size of the large objective lens; the eyepiece of the instrument merely magnifies the image formed by the objective and, consequently, it is useless to increase the power of the eyepiece beyond the point where the spurious diffraction disks become noticeable. The formula for determining resolving power is

$$\text{Minimum angle resolved} = \frac{5 \text{ sec. of arc}}{\text{Aperture in inches}};$$

in other words, the Yerkes 40-in. telescope gives to a point of light a fictitious angular size of 0″.12. These diffraction patterns are caused by the nature of light itself, and, being therefore unavoidable, we can only decrease the size of the spurious disk by increasing the aperture of the telescope.

But increasing the size of a telescope introduces great engineering difficulties. It is doubtful whether *refracting* telescopes, i.e., those in which the light is collected by a huge objective lens, will ever be constructed appreciably larger than the 40-in. refractor of the Yerkes Observatory, for lenses greater than 4 or 5 ft. would probably bend under their own weight and the resulting image become worthless. Then, again, the glass employed for the lenses must be absolutely homogeneous throughout and free from defects. For these reasons, telescopes of the larger apertures are all of the *reflector* type—an invention we owe to Newton. In the Newtonian form, a mirror is situated at the base of the telescope and reflects the light back up the tube so that it converges on a small plane mirror which, in turn, reflects the rays to an eyepiece at the side of the tube. In looking at a star with a Newtonian reflector, the line of sight is therefore at right angles to the direction of the object.

THE change in fashion in telescopes is interesting. From Galileo to Newton, the only type was the refractor, then from 1670 until the beginning of the nineteenth century the reflector took the lead. Improvement in the refracting telescope through the use of compound objective lenses caused the reflector to be almost displaced by the refractor, but, with the demand for greater and still greater telescopes in the last twenty years, mirrors are again in favor. The largest reflector now in operation is the 100-in. Hooker reflector of Mount Wilson Observatory; the 100-in. mirror alone weighs 4 tons and the moving parts of the telescope 100 tons. Meanwhile we are kept informed of the progress on that stupendous undertaking, the 200-in. reflector.

The question naturally arises as to how large a telescope must be in order to show us the detailed structure of a star similar in actual size to the sun. Let us imagine ourselves as situated 260,000 astronomical units from our sun, a distance a little less than that separating the earth and the nearest known star Proxima Centauri. At 1 astronomical unit, the sun has an angular diameter of $32'$; consequently at 260,000 astronomical units we compute that its apparent diameter would be $0''.007$. Suppose that, at this distance, we studied our sun with a 200-in. telescope. Inasmuch as the formula pertaining to resolving power informs us that, with such an aperture, the image of a point of light would be a spot $0''.025$ in diameter, we conclude that the shape of the solar image would be essentially spurious. Now, if we had a telescope of 4,800-in. aperture, so that this spot would shrink to $0''.001$, there might be some purpose in studying the stellar image itself.

OUR example is not to be regarded as propaganda for a 400-ft. mirror because, for maximum efficiency, such a hypothetical instrument would require to be mounted high in the stratosphere. Before entering our telescopes, the light of a star is at the mercy of air currents of variable density, and the larger the aperture the wider the air-path and the greater the chance of distortion. We say a star twinkles because of our atmosphere and attempt to distinguish between stars and planets in this fashion, the steadier light of the planets being accounted for by their appreciable angular diameter. For example, Jupiter has an average angular size of 40″ of arc, while that of the stars is believed not to exceed 0″.05. More than half-a-million stars would have to be placed together to occupy the same area as Jupiter in our sky, and we would expect the combined light of such an aggregation of twinkling stars to be fairly steady. But even planets twinkle if they are close to the horizon or when atmospheric disturbances are more pronounced. The astronomer calls this condition "bad seeing," for then the telescopic image of even a planet becomes blurred and visual observations are valueless.

Despite the absence of a true image, a wealth of information may be derived from starlight. In the Yerkes 40-in. refractor, as well as the other instruments mentioned, a clock drives the telescope about an axis parallel to that of the earth so that, when once directed at a star, the telescope automatically follows the object. To arouse interest in a star's image, we might stop or disconnect the clock-drive. The stellar points are then seen to pass rapidly across the field and, with the Yerkes 40-in., their apparent motion may easily be increased a thousand fold. We begin to realize how stellar motions, which would take thousands of years to be discernible with the unaided eye, become quite noticeable in a year or so with such an instrument, and why it is that, during the last two centuries, we have begun to regard the term "fixed stars" as a misnomer.

IN CHAPTER 1 we considered annual parallax (the yearly motion of the nearby stars with respect to the more remote ones) as a proof of the earth's revolution. Of much more importance to astronomers is the fact that parallax is a means of directly establishing the distance of the nearby stars. By the ANGLE OF PARALLAX, or simply the parallax, of a star is meant the maximum angular displacement of the star due to the orbital motion of the earth. (Because of the earth's elliptical orbit, this definition is not absolutely correct. If the earth moved in a circle 1 astronomical unit in radius, then the angle of parallax would be as previously defined.) As seen

THE SIDEREAL UNIVERSE

from the star, the earth would describe an ellipse about the sun, and the maximum angular separation of the earth and the sun would be equal to the parallax of the star. Incidentally, with the instruments we now possess, it would be impossible to detect the earth or any of the planets from even the nearest star, though our sun would be clearly discernible as a first-magnitude star.

Parallactic determination of a distance is analogous to the method of finding the distance across a stream or to the moon (see chap. 4). A base line is always required; in surveying a terrestrial one is employed, while for the nearby stars we utilize two orbital positions of the earth of known distance apart.

To establish the proposition that a parallax of $1''$ implies a distance of 206,265 times the mean distance between earth and sun involves only elementary geometry. As a convenient yardstick for celestial distances, we have the PARSEC, defined as follows:

$$1 \text{ parsec} = 206{,}265 \text{ astronomical units},$$

so that a parallax of $1''$ implies a distance of 1 parsec. Expressed in miles,

$$1 \text{ parsec} = 19{,}160{,}000{,}000{,}000 \text{ mi.}$$

Now in one year light travels nearly 6,000,000,000,000 mi.—a distance we call the LIGHT-YEAR and, hence, by simple division, we conclude that it takes light 3.26 years to travel 1 parsec. We express this result by saying that

$$1 \text{ parsec} = 3.26 \text{ light-years}.$$

This formula enables us to convert parsecs into light-years, the latter unit conveying a better idea of the immensity of the distance involved. The greater the distance of a star, the smaller its parallax; in fact,

$$\text{Distance in parsecs} = \frac{1}{\text{Parallax in seconds of arc}},$$

and, since the largest parallax discovered is $0''.785$, we conclude that the nearest star found to date is 1.27 parsecs or 4.15 light-years, away.

KNOWING the distance of a star and its apparent magnitude, i.e., its magnitude as seen from the earth, it is possible to deduce its actual or intrinsic brightness. The latter may be expressed in terms of the magnitude a star would have at some standard distance, say 10 parsecs, the ABSOLUTE MAGNITUDE as it is called. When we deal with absolute magnitude, we are placing all the stars at a distance of 10 parsecs, or 32.6 light-years, and comparing their brightness accordingly. At this distance the sun would appear as a fifth-magnitude star; or, more precisely,

$$\text{Absolute magnitude of sun} = 4.85 \ .$$

Up to the present, only ten stars have been found within 11 light-years of the earth, and their parallax, distance, apparent and absolute magnitudes, are given in the following table:

Name	Apparent Magnitude	Parallax	Distance in Light-Years	Absolute Magnitude	Luminosity $\odot = 1$	Proper Motion in 1 Yr.	Spectral Type
Proxima Centauri......	10.5	0."785	4.2	15.0	0.0001	3."9	M
α Centauri*...........	{ 0.3 1.7 }	.76	4.3	{ 4.7 6.1 }	{ 1.1 0.2 }	3.7	{ G K
Barnard's star........	9.7	.56	6.0	13.4	0.0004	10.2	M
Lalande 21185........	7.6	.39	8.3	10.6	0.005	4.8	M
Sirius.................	− 1.6	.38	8.6	1.3	26.0	1.3	A
(Anonymous)..........	12.0	.34	9.6	14.7	0.0001	2.7	....
Procyon...............	0.5	.31	10.4	3.0	5.6	1.2	F
ε Eridani.............	3.8	.30	10.7	6.2	0.28	1.0	K
61 Cygni*............	{ 5.6 6.3 }	.30	10.9	{ 8.0 8.7 }	{ 0.06 0.03 }	5.2	{ K K
τ Ceti................	3.6	0.30	10.9	6.0	0.35	2.0	K

* Binary.

TABLE 8

STATISTICS OF THE NEAREST STARS

There is little doubt but that, as time goes on, additions will be made to this table. A star as faint as Anonymous might still be within 11 light-years and its proximity to the earth pass unnoticed, for there are more than a million stars brighter than magnitude 12. The complete list of stars for which annual parallax has been observed totals a few thousands—a paltry number when compared to the thousands of millions of stars within the reach of our telescopes. This is no reflection upon the diligence of astronomers, for not quite

THE SIDEREAL UNIVERSE

a century has passed since Henderson in 1840 succeeded in determining the first stellar parallax—that of α Centauri. Insufficient time, however, is not the major cause; the vast majority of stars is so far away that present-day instruments are incapable of detecting their annual parallax. From statistical studies, it is found that the measured angle of parallax is likely to be in error by $0\rlap{.}''003$ (the angular size of a dime situated in New York and viewed from Chicago) and, because of this small error, a parallax of less than $0\rlap{.}''01$ can hardly be regarded as indicative of the star's actual distance. Consequently, for reasonable accuracy, direct determination of distance by annual parallax is limited to about 300 light-years.

THE information derived from stars of known parallax is treasured by the astronomer. It is his "Rosetta Stone"—the key to the nature of stars whose light has traveled for thousands or even millions of years to be finally intercepted by his telescope. By ingenious statistical methods the peculiarities of these nearby stars are utilized to find the extent and structure of the stellar universe. We therefore pause and study what little information has been included in Table 8.

In our exposition of the parallactic proof of the revolution of the earth, we noted that in general such measurements are complicated by stellar motion relative to the sun and, if we viewed the stars for a few hundred years, a change in the angular position of certain ones would be observed. As seen from the earth, this PROPER MOTION is combined with the annual parallactic alterations of a star's apparent position and, in Table 8, we list its yearly amount in seconds of arc for the nearby stars. For example, Barnard's star moves $10\rlap{.}''2$ per year, so that, in 352 years, this star will have shifted its position 1°.

We hasten to mollify the reader who has mastered the methods of chapter 3 for estimating the position of "fixed" stars by assuring him that Barnard's star is exceptional, for it heads the list of stars with greatest proper motion. In fact, all the stars in Table 8 rank high in proper motion and, of the approximately twenty stars with proper motions greater than $3''$ per year, five occur in this table.

Let us draw an analogy between the motion of a star and the flight of an airplane. As a rule, the closer the airplane, the more rapidly it appears to change its position in the sky; in like manner, the proper motion of a star depends not only on its actual velocity relative to the sun, but also upon its distance and direction of motion, and we attribute the high proper motion of nearby stars to their proximity rather than to their actual velocity. Conversely,

Name	Distance in Light-Years	Proper Motion in 1 Yr.	Absolute Magnitude	Luminosity $\odot = 1$	Spectral Type	Distance in Light-Years $\times$ Proper Motion
α Centauri*	4.3	3".7	{ 4.7 6.1	1.1 0.2	G } K }	16
Sirius	8.6	1.3	1.3	26.0	A	11
Procyon	10.4	1.2	3.0	5.6	F	12
Altair	16.0	0.66	2.4	9.2	A	11
Fomalhaut	24	0.14	2.0	13.5	A	3
Vega	26	0.34	0.6	50	A	9
Pollux	32	0.62	1.2	28	K	20
Arcturus	40	2.29	−0.2	100	K	90
Capella	50	0.43	−0.6	150	G	22
Regulus	60	0.24	0.2	70	B	14
Aldebaran	60	0.20	−0.1	90	K	12
Achernar	70	0.09	−0.9	200	B	6
Betelgeuse	200	0.03	−2.9	1,200	M	6
Spica	200	0.05	−3.1	1,500	B	10
α Crucis*	200	0.05	{ −2.7 −2.2	1,000 650	B } B }	10
β Centauri	300	0.04	−3.9	3,100	B	12
Antares	400	0.03	−4.0	3,400	M	12
Rigel	600 ?	0.005	−6.0 ?	18,000 ?	B	3 ?
Canopus	700 ?	0.02	−7.0 ?	80,000 ?	F	14 ?
Deneb	700 ?	0.004	−5.0 ?	10,000 ?	A	3 ?

* Binary.

TABLE 9

STATISTICS OF THE FIRST-MAGNITUDE STARS

we expect stars of low proper motion to be relatively far away. Let us check this in the statistics derived for the twenty first-magnitude stars. In Table 9 these stars are arranged according to distance.

OF THE first-magnitude stars visible in the middle-northern latitudes, Arcturus and Sirius have the greatest apparent motion and, as might be anticipated, these two stars were the first ones noticed to have motions of their own. Halley in 1718 called

THE SIDEREAL UNIVERSE

attention to the fact that Arcturus and Sirius had moved southward since the time of Ptolemy by about 1° and $\frac{1}{2}$°, respectively. In the third column of Table 9, it is seen that as a rule the proper motion is less for the more remote stars. The notable exception is Arcturus; its large proper motion cannot be accounted for wholly on the basis of its proximity. We conclude that the actual velocity of this star must be exceptionally high—a conclusion that has been verified by other types of measurements.

According to the table, Betelgeuse, Spica, α Crucis, β Centauri, and Rigel are approximately ten times as far away from us as Altair, Fomalhaut, Vega, Pollux, and Capella; furthermore, on the whole the former group have proper motions about one-tenth as great as those of the latter. We therefore deduce that the actual velocities of both groups are roughly the same. In order to point more clearly this dependence of proper motion on distance, we have listed in the last column of Table 9 the product of distance-in-light-years and proper-motion-in-seconds-per-year. Each entry in this column may be regarded as representing the proper motion the star would have if the distance between the solar system and the star were diminished to 1 light-year, preserving, however, present velocities and direction of motion. The important aspect of this product is its approximate constancy and its independence of distance. Of the twenty entries, ten have values between 10 and 14 inclusive and these ten are distributed throughout the table. Consequently, our statistical study of the table leads us to the conclusion that, whatever the distance of the star, there is an even chance that

(Distance in light-years) × (Proper motion in seconds per year) =
A number between 10 and 14 .

Given merely the proper motion of a star, we may utilize these results to derive some notion of its probable distance. To illustrate, suppose the motion of a star is 0".002 per year. According to our statistical formula, the probability is one-half that the distance of this star is between 5,000 and 7,000 light-years. Though no intelligent person would attempt to plan his life on the basis of mortality tables, nevertheless life insurance companies formulate their policy with these tables in mind. Similarly, statistical inferences should not be applied to particular stars but rather to aggregates. After all, in dealing with a few billion stars the astronomer should be permitted to adopt an attitude analogous to that of the actuary, i.e., to consider the individual only as a member of a group.

THE foregoing was merely to show that a method for deriving distance from proper motion is feasible, and it is not to be regarded as illustrative of the actual procedure followed in stellar statistics. As a matter of fact, we should not even try to generalize on the basis of only twenty stars; in stellar statistics all available data for thousands of stars are collected and painstakingly analyzed. Not one but many methods have been developed for obtaining estimates of stellar distances, and the results obtained by means of one scheme are checked with those of another. Proper motion alone does not carry us far enough in our exploration of the stars. Though it is true that any star will be observed to change its position in a sufficiently lengthy interval of time, even the most dispassionate astronomers are likely to become a trifle impatient while trying to measure the proper motion of a star 50,000 light-years away. They carefully record its apparent position for the benefit of future observers and then examine the light of the star itself for clues as to its distance and physical properties.

Apparent brightness of stars and proper motion are alike in one respect—they both diminish with distance. As a consequence, many of our conclusions regarding the extent of the visible universe are derived from observations of apparent luminosity. The physicist has found that the intensity of light varies inversely as the square of the distance of the source, provided there is no intervening matter to absorb the light. For example, if there were no absorption of light in interstellar space, a star at twice its distance would appear one-fourth as bright.

It is this *proviso*—no absorption—that has shaken the confidence of some astronomers in the reliability of results based on apparent magnitude alone. There is abundant evidence of the presence of obscuring matter in space; for instance, the dark lanes in the Milky Way indicate a screen of material that is cutting off the light of more remote stars. Then, again, in 1904 J. Hartmann showed that certain colors in starlight were diminished in intensity, because of absorption by clouds of calcium and perhaps sodium. Recent studies of O. Struve indicate that these clouds pervade all space and, from theoretical considerations, A. S. Eddington concludes that their density is extremely low, the total mass in a sphere of radius equal to that of the earth being of the order of a pound.

The absorption of light by interstellar calcium or sodium is not believed to alter appreciably the total light of a star. It affects only certain pure colors and, inasmuch as starlight is a composite of many colors, the diminution of the intensity of a relative few would not noticeably reduce the combined light of the aggregate. But there are other ways in which the luminosity of a star may suffer. Just as the earth's atmosphere scatters the light of the sun, so a dust of minute particles in space would alter starlight. When close to the horizon, the sun's rays reaching us pass through a greater thickness of atmosphere, the scattering of blue light is more pronounced, and the sun has a decidedly reddish tinge. Now if dust is present in space, its effect will be more noticeable on the light of distant stars and, consequently, the discovery that the more remote stars are on the whole a trifle redder than the nearby stars is regarded as almost conclusive proof of the presence of celestial dust. If this be so, the scattered starlight should brighten the entire night sky—a phenomenon that is actually observed. Part of this illumination, however, is traceable to the zodiacal light, *gegenschein*, and atmospheric aurorae, but the remainder may be due to scattering of starlight. Perhaps the sporadic meteors, which have been shown to be of cosmic origin, are the larger particles of this interstellar matter that is being swept up by the earth.

THE precise magnitude of this absorption and scattering has not as yet been determined, though the consensus of opinion is that assigned stellar distances, obtained by disregarding cosmic dust, will not be radically changed. Be that as it may, for the present we postulate that interstellar space is more or less transparent. With this hypothesis it is a relatively simple task to estimate how bright our sun would be at various distances.

Distance in Light-Years	Magnitude
0.000016	−26.7
1	− 2.7
10	2.3
100	7.3
1,000	12.3
10,000	17.3
100,000	22.3
1,000,000	27.3

MAGNITUDE OF SUN AT DIFFERENT DISTANCES

If the stars, including our sun, were all of equal intrinsic brightness, this table would enable us to compute their distances, given only their apparent magnitudes. For example, we could then say that all stars visible to the naked eye were within 100 light-years of the earth. A glance at either Table 8 or 9, however, reveals that the assumption of equal luminosity is untenable; our nearest known star, Proxima Centauri, is not even a millionth as bright as Antares. Furthermore, most of the stars contained in Table 8 are of low luminosity relative to our sun, and are known as DWARF stars. Since there is no reason to believe that our neighbors are exceptional, we conclude that the dwarfs constitute a large percentage of the stars in the universe. Out of the eighteen stars known to have a parallax greater than 0".25, more than half are dwarfs with intrinsic luminosities ranging from 1/100 to 1/10,000 that of our sun, while only less than 2 per cent are brighter. In Table 9 we find stars thousands of times brighter than the sun, the GIANT stars. Though a large percentage of the first-magnitude stars are giants, nevertheless they are greatly outnumbered by the dwarfs. On the basis of the number of dwarfs in the neighborhood of the sun we deduce that, in the volume of space occupied by the twenty-first magnitude stars, there will be tens of thousands of dwarfs.

This classification of certain stars into dwarfs and giants is based solely on luminosity. Now great luminosity does not necessarily imply immensity, e.g., an electric arc may be brighter than a bonfire. The intrinsic brightness of a star depends not only on the area of the radiating surface but also on the intensity per unit area. If giants and dwarfs radiated the same amount of light per unit area, then it would follow that this division is also one of size. A digression, however, is necessary before we follow this lead.

THE SIDEREAL UNIVERSE

IN PAST chapters we have had occasion to mention the spectroscope—an instrument for resolving light into its component colors. Two and a half centuries ago Newton passed a ray of sunlight through a glass prism and refracted it into a rainbow of colors—the spectrum. Thus began the science of spectroscopy. It is only during the past century that the spectroscope has been applied to the stars, but, in this period, it has completely revolutionized observational procedure and has become the most important telescopic accessory. The light of the star is concentrated by the objective lens on to the slit of the spectroscope, where it is then broken up by a succession of prisms into its spectrum, and this band of colors is photographed to give a record of the relative intensities of the component colors in the star's light. That the spectra of stars differ greatly is evident from the four illustrations here reproduced.

In the description of the sun, we pointed out that elements in the reversing layer absorbed certain colors of the photosphere.

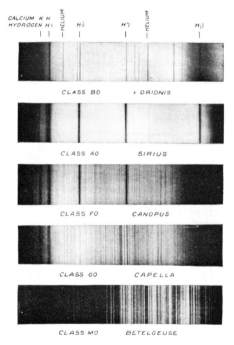

PLATE 21 Representative Stellar Spectra of the Principal Classes

Consequently, when we spread out the sunlight into its spectrum, we find that it is a continuous band of rainbow colors crossed by dark lines, corresponding to the pure colors removed by the reversing layer. These dark lines are called FRAUNHOFER LINES after the physicist who first mapped and discussed them in 1817, and in 1895 H. A. Rowland published a catalogue of the position and intensities of 16,000 of these lines in the solar spectrum.

IN GENERAL, the spectra of the stars resemble that of our sun, i.e., a continuous band of colored light crossed by dark absorption lines. A small percentage of the stars have bright emission lines superimposed on the band. This is due to the radiation of light by a hot gas at low pressure, and the emitted pure colors are precisely the same ones this gas would absorb from light passing through it. But the numbers and intensities of such emission lines differ greatly in stellar spectra. In 1864 Secchi divided stellar spectra into four broad classes, but this classification has been largely superseded by that of the *Draper Catalogue* of the Harvard Observatory. In this scheme capital letters are used to denote main divisions, and we are indebted to Miss Annie Cannon for the classification of about a quarter of a million stars. We shall here omit the many subdivisions denoted by small capitals and numbers and only outline very briefly the principal categories.

Type	B	A	F	G	K	M
Distinguishing characteristics of lines	Lines of hydrogen, neutral helium, ionized metals	Strong hydrogen lines, ionized metals	Lines of neutral metals present	Metallic lines strong	Lines of titanium oxide present	Strong titanium oxide bands
Color	Blue or bluish-white	White	Yellowish-white	Yellow	Red	Red
Temperature	20,000° C.	10,000° C.	8,000° C.	6,000° C.	4,000° C.	3,000° C.
Typical example		Sirius	Procyon	Sun	Arcturus	Antares
Radiation per unit area (Sun=1)	100	8	2.5	1	0.2	0.02

TABLE 10

PROPERTIES OF THE MORE COMMON SPECTRAL TYPES

THE SIDEREAL UNIVERSE

In the order of simplicity of absorption lines, the spectral classes are O, B, A, F, G, K, M, R, N, S, and this sequence also corresponds to a progressive change in the color of the star from blue or bluish-white to red. The color of a star is in turn an indication of its surface temperature; consequently, given spectral type, the approximate temperature and rate of radiation are known. The dependence of color on temperature may be illustrated by heating a bar of iron; with increase in temperature the bar changes as follows: a dull red, yellow, yellowish-white, white and bluish-white.

WE NOW return to the dwarfs and giants and apply our knowledge of spectral type in an attempt to estimate their size. The last row of the Table 10 gives the approximate luminosity a star would have if it were the same diameter as our sun; e.g., on this basis a type B star would radiate twenty times as much light as does the sun. On studying this last line we see that the enormous variation in absolute luminosity from dwarfs to giants cannot be wholly attributed to brightness per unit area and, consequently, we may regard these terms as indicating size. In fact, Russell showed in 1913 that this division into dwarf and giant stars holds in all spectral classes, though it is most obvious in the redder stars. Some of these giants are of so high a luminosity that they are fittingly termed SUPERGIANTS.

Let us consider the supergiant Antares. According to the table, its rate of radiation per unit area is about 1/50 that of the sun but its total radiation is 3,000 times greater. We deduce that the radiating surface of Antares is roughly 150,000 times the solar one, the diameter 400 times. Fortunately it has been possible to check this result by direct measurements. The actual diameter here estimated corresponds, at a distance of 300 light-years, to an apparent diameter of 0".04. Though present-day telescopes are too small to reproduce faithfully a disk of this angular size, the effective aperture for measuring stellar diameters may be increased by reflecting the starlight into the telescope from two mirrors, placed at a considerable distance apart (a modified form of the interferometer).

In 1920 the physicist Michelson brought this method of measuring stellar diameters to the attention of astronomers and, with the aid of the Mount Wilson observers, mounted two mirrors 20 ft. apart on the end of the 100-in. reflector. With such equipment it is possible to measure illuminated disks greater than 0".02 of arc. (The image, however, is not that of the star. It consists of a series of fine bright and dark fringes, and it is the behavior of this interference pattern that permits the measurement of the star's diameter.) The diameters of stars obtained in this manner by F. G. Pease and J. A. Anderson are:

(Diameter of Sun = 1)

Antares	428
α Herculis	403
Betelgeuse	296
Mira Ceti	150
β Pegasi	140
Aldebaran	39

We conclude that the volume of Antares is 70,000,000 times that of the sun.

The agreement between the theoretical estimates of Eddington and Russell and the observed stellar diameters of the giants strengthens our confidence in the computed diameters obtained for dwarfs, based solely on spectral type and absolute luminosity. One particular dwarf has received considerable attention—the companion of Sirius. In 1862 Alvan G. Clark discovered that a faint star (magnitude 8.4) revolved about the brilliant first-magnitude Sirius, i.e., Sirius is a binary star. Further study has shown that their mean distance apart is 1,800,000,000 mi. and that their period of revolution is 48.8 years. Following the methods of celestial mechanics (chap. 5), their combined mass has been found to be 3.4 times that of the sun, while the mass of the faint companion is about the same as the sun. But this companion star is a dwarf and computations assign to it a diameter 1/30 that of the sun, a volume 1/30,000 and, consequently, a mean density 50,000 times that of water. A spoonful of matter of this density on the earth would weigh a ton!

WHAT has been said of the companion of Sirius seems to apply to all the white dwarfs. Despite the fact that the stars vary enormously in luminosity, the range of stellar mass is limited. Our direct knowledge of this is, of course, derived from binary stars and, in a manner similar to that utilized in determining the mass of the planet Mars from the period of revolution of its satellites, we are able to compute the combined masses of two stars revolving about each other. The available data are not limited to TELESCOPIC BINARIES (twin stars whose components are resolvable by means of a telescope), but also include SPECTROSCOPIC BINARIES.

With the spectroscope it is possible to determine the actual rate at which a star's distance from the earth is increasing or decreasing. Light consists of waves and color depends on the frequency of those waves, just as the pitch of a musical tone depends on the number of vibrations per second. Now if our distance from a source of illumination is diminishing, we will encounter more waves per second and the color will change. Theoretically, if we could approach a red light with a velocity of 150,000 mi. per second, the light would appear blue. Terrestrial motions, however, are not sufficiently high to detect these DOPPLER EFFECTS, as they are called, but they are noticeable in the lines of stellar spectra. If the star is approaching there is a slight shift to the blue, if receding the shift is to the red and, from this shift, we are able to compute its velocity of approach or recession.

When two stars are close together and rapidly revolving about each other, each star is seen through its spectrum to approach and recede periodically by a corresponding slight back-and-forth motion in its spectral lines. Spectroscopic binaries identified in this manner have small periods ranging from 8 hr. to several thousand days. Though a few long period spectroscopic binaries are also resolvable into two stars by a telescope, the majority of stars identified as telescopic binaries have periods ranging from a few years to several thousand years. The number of these spectroscopic binaries now totals 1,000 and is rapidly increasing; furthermore, it is suspected that at least one star in every five is in this category. As for the telescopic binaries, in 1827 F. G. Struve published a catalogue of 3,110 such pairs, and now R. G. Aitken's catalogue of 1931 tabulates information concerning 17,910. Further investigation discloses the fact that some of the supposed binaries actually are MULTIPLE STARS; e.g., Castor is really a system of six stars.

The ECLIPSE VARIABLES fall in a particular subclass of the spectroscopic binaries. Algol (the Demon Star) is of this type, as well as β Lyrae. The two components of the binary are so close that the one star eclipses the other. In Algol the one component is considerably fainter than the other, so that when the faint one eclipses the brighter one, there is a decided drop in apparent magnitude. The two stars of β Lyrae are of approximately the same brilliancy and, therefore, the variations in apparent intensity are not so marked. From the study of such stars, information is derived as to their shape and density, the latter being in general very low, sometimes 1/2,000,000 that of the sun.

WE SUMMARIZE the salient results obtained from mass determinations of binary stars. On the whole, the masses of stars are comparable with the sun and we do not expect to find stars with masses greater than 100 times that of the sun. This result was also obtained by Eddington from theoretical considerations of stellar interiors and, in 1924, he pointed out this striking relation: whatever the spectral type of a star, its intrinsic brightness seems wholly dependent upon its mass, increasing luminosity corresponding to increasing mass. This mass-luminosity relation is tabulated as follows:

Mass (Sun=1)	12	4	1	1/3
Absolute magnitude	−2.5	0.0	5.0	10.0

On the basis of this table we conclude that such a giant star as Betelgeuse (absolute magnitude −2.9), with a volume more than 2,000,000 times the sun, has a mass only about 15 times the solar one.

IT IS excusable to speculate as to whether such gigantic stars of low density might collapse or whether, on the other hand, those white dwarfs with their incomprehensible density might explode and, when we do see a star suddenly blaze forth and radiate 10,000 times more than its normal amount of heat and light, we suspect that some such catastrophe has actually occurred. We refer to the NOVAE or temporary stars, similar to the one that leaped into prominence and became for a time a second-magnitude star—Nova Herculis 1934. The brightest of these novae on record are Nova Cassiopeiae, which was discovered by Tycho Brahe in 1572 and attained magnitude −4, i.e., brighter than the planet Venus, and Nova Ophiuchi, discovered by Kepler in 1604, which reached magnitude −2. The next brightest occurred in recent years: Nova Persei in 1901 (maximum magnitude 0.0) and Nova Aquilae in 1918 (maximum magnitude −0.5). Unfortunately these novae are

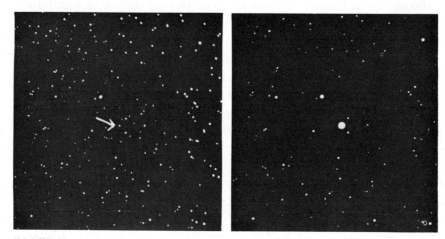

PLATE 22 Nova Aquilae of 1918 before and after Outburst

not usually noticed before they burst forth, and our study of them commences when they are about at maximum but, where information is available, it is found that the outburst of light is sudden: for instance, Novae Aquilae changed from an eleventh-magnitude to a first-magnitude star within two days, increasing its rate of radiation fifty thousand fold. A rapid decline generally follows, gradually diminishing in rate until in a few months, or years as the case may be, the star is not detectable or is back to the magnitude from which it sprang, though there may be secondary outbursts of lesser brilliancy.

Direct observation indicates that there is an actual outburst of material and, in the case of Nova Pictoris (1925), Spencer Jones found spectroscopically that, before reaching its maximum, the radius of the nova increased daily by 14 times the radius of the sun; and, after the explosion, a greenish nebulous shell was telescopically seen to surround the nova, this envelope continuing to increase at the rate of 2″ per year. Knowing the actual speed at which the material is ejected and its apparent angular change, we can determine the distance of the nova. In this manner the distance of Nova Pictoris has been estimated at about 20,000 light-years and its absolute magnitude at maximum −8, equivalent in radiation to 160,000 of our suns. On the basis of the meager data available, it has been concluded that there is one nova per year per hundred million or billion stars.

We do not know the ultimate fate of the novae. Some claim that the gaseous envelope is dissipated in space and that the central star becomes a white dwarf; others suggest that once a nova always a nova and that, after a sufficient interval, it will repeat its performance. Perhaps the novae are VARIABLE STARS with times of maximum brilliancy separated by tens of thousands of years.[1]

IN CHAPTER 3 we mentioned the long-period variable Mira, which attains its greatest brightness after intervals of a little less than a year. We could list, but will not, many other variables with periods ranging from 11 to 450 days and fluctuations in brightness of 8 magnitudes, as well as stars that have no apparent regularity in their behavior. We must, however, mention the CEPHEID VARIABLES, because information concerning this group has proved so useful in estimating stellar distances. They are named after δ Cephei, the typical representative star, but are not confined to this particular constellation. They fluctuate in brightness by 1 magnitude in a period ranging from a day to one or two months, and their variations in intensity are altogether different from the eclipsing binaries. The theory of their behavior is that they are *pulsating*, i.e., contracting and then expanding to produce a corresponding fluctuation in brilliancy. A study of Cepheid variables reveals that their actual luminosity is definitely related to their period and, consequently, knowing the interval between maximum

---

[1] Contradicting this suggestion, we note that two "new stars" seem to have been ejected from the core of Nova Pictoris, while Nova Herculis appears to have suffered a bifurcation into two stars!

intensities for a star of this type, absolute magnitude is determinable. When this knowledge is combined with brightness as seen from the earth, we are able to deduce the distance of the star. It is in this manner, by employing giant Cepheid variables, that we obtain distances of groups of stars hundreds of thousands or millions of light-years away.

So far we have been dealing with stars—if we care to classify all these objects under one such head. We have had occasion to mention that there was matter scattered in space and, in certain regions, we might expect local condensations of this matter cutting off the light of the more remote stars. Then, again, brilliant stars may be close to these gaseous clouds, pouring energy into them and causing them to radiate light of their own. We attempt in this way to account for the DARK and BRIGHT NEBULAE in space.

IN THE constellation of Orion, right in the Sword, is a star that appears "fuzzy" to the eye, but when photographed through a large reflector, exposing the plate for two hours or so, a luminous gaseous cloud is revealed with a wealth of detail. The Great Nebula in Orion, as it is called, is believed to be 600 light-years away and covers such an enormous volume of space that it takes light 10 years to pass across it. Recent spectroscopic observations indicate that this huge, apparently irregular, gaseous cloud is rotating in a period of several hundred thousand years, and it has been inferred that its mass is 10,000 times that of the sun, its density one-million-billionth (1/1,000,000,000,000,000) that of the air we breathe. These results are, of course, subject to question, but, nevertheless, this nebula must be extremely tenuous, otherwise, its attraction would appreciably alter the course of the stars.

The Great Nebula in Orion is by no means the only one of its type. This class of DIFFUSE NEBULAE has many shapes and a variety of picturesque names: the Dumbbell Nebula, the Crab Nebula, the Keyhole Nebula, etc., and, in the constellation of Cygnus, there is the North America Nebula, so called because of its resemblance to the shape of our continent.

Photographic studies of the Pleiades show that this group of stars is surrounded by vast gaseous clouds that look like dabs of

PLATE 23 THE GREAT GASEOUS NEBULA IN ORION, SHOWING ADJACENT NEBULA AND STARS

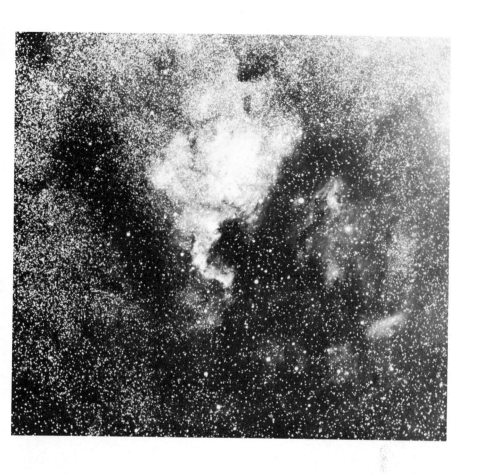

PLATE 24 THE NORTH AMERICA NEBULA IN CYGNUS, GASEOUS

PLATE 25 NEBULOSITY IN THE PLEIADES EXPOSURE 3 HR. 30 MIN. TAKEN WITH 2-FOOT REFLECTOR BY RITCHIE

THE SIDEREAL UNIVERSE

paint. These nebulae are definitely associated with stars, and Edwin Hubble has shown that this is true of the other brighter nebulae. Again, on a background of stars, we have black patches, the dark nebulae, many of which have outlines closely resembling certain identified bright nebulae. We conclude that the bright and dark nebulae are materially the same, but the former have nearby stars to supply illumination while the latter cannot shine by themselves.

PLATE 26 The "Ring" Nebula in Lyra, Gaseous

In addition to the diffuse nebulae, there is a relatively small number (about 150) of round- and oval-shaped nebulae, known as the PLANETARY NEBULAE. This is a misnomer, for, though they appear through a small telescope to be somewhat like a planetary disk, in reality they are nebulous shells or spheres, probably with diameters thousands of astronomical units. Many of them are brighter at their circumferences, which gives them a ringlike appearance such as is seen in the famous Ring Nebula in Lyra. In the center of certain ones is a faint star, and it has been suggested that all planetary nebulae owe their light to a star, or stars, in their interior.

THE objects we have been describing are found mostly in the plane of the Milky Way. Now it is in this hazy band of light that the greatest concentration of stars appears to be, but, since statistical studies reveal that by and large the stars are not crowded together any closer here than in any other region of space, we conclude that in the plane of the Milky Way the stars extend to greater distances. From an exhaustive study of observed data, Trumpler deduced that our sun is in an aggregate of thousands of millions of stars with enormous interposing nebulae, the whole being distributed in the form of a huge disk—the GALAXY. This galactic disk he claims is 3,000 light-years in thickness and its maximum diameter 30,000 light-years, while our sun is near the center of the disk. The stars are not distributed uniformly in the galaxy but are mostly concentrated into swarms or clouds of stars. Our sun itself is believed to belong to a local cloud which, along with other star-clouds, forms this enormous galaxy. Spectroscopic observations show that, relative to this cloud, the sun with its planets is moving through space in a straight line, with a velocity of 12 mi. per sec., toward a point in the sky that may be defined by right ascension 18 hr., declination 30°, i.e., toward the constellation of Hercules. This solar movement is noticed in the proper motion of the stars, for stars directly in the path of the sun seem on the whole to separate in order to make room, while those stars left behind seem to gather together.

OFF the plane of the Milky Way and at distances of 20,000 to 200,000 light-years (obtained from Cepheid variables and other studies of stellar luminosity) are the GLOBULAR CLUSTERS, of which the Great Cluster in Hercules is a typical example. The stars in this group are distributed in a sphere of diameter about 200 light-years, the number of stars per unit volume being greatest at the center. Though only 40,000 star images have been counted on photographs, inasmuch as only giants would be visible at such dis-

PLATE 27 GREAT GLOBULAR CLUSTER IN HERCULES, MESSIER 13

tances, it is not unlikely that the total number in the Cluster of Hercules is several millions. In all, less than a hundred of these clusters have been found, mostly situated on one side of the sky. Whereas Trumpler regards them as detached from the galaxy, Shapley considers the clusters an integral part of the organization.

Taken together they form a system the center of which lies in the Milky Way in the direction of Sagittarius, at a distance Shapley estimates as 52,000 light-years. Neglecting a few isolated clusters, the outline of this system is similar to that of a double-convex lens with greatest thickness 30,000–40,000 light-years and maximum diameter 240,000–300,000 light-years. If we adopt Shapley's assumption that the globular clusters mark the boundaries of the galaxy, then it follows that these dimensions are those of our galaxy, making it a much greater structure than that deduced by Trumpler. Now, if Shapley's picture is the correct one, our sun is 52,000 light-years away from the center and about midway between the borders of the galaxy. As confirmation of this eccentric solar position, we note that the densest star-clouds of the Milky Way occur in and near Sagittarius—the location of the hypothetical center—and here also the novae are most numerous. In addition to this, Plaskett's recent investigations show that the galaxy itself is rotating and that the distance to the center of rotation is about 40,000 light-years, essentially the same as the center of Shapley's conception of the galaxy. The period of rotation is 250,000,000 years and, as the age of the earth is several billion years, it follows that the galaxy must have made several gyrations during the life of the Solar System.

One objection to including the globular clusters in the galaxy is that no such symmetrical structure of stars is found in the Milky Way. This, however, does not preclude the possibility of the presence of such clusters there. The hosts of stars could render them indistinguishable or they could be occluded by dark nebulae. Near the plane of the Milky Way, however, there are compact groups of stars traveling with identical velocities in a straight line. In all, some 250 of these so-called OPEN CLUSTERS have been found, among which are the famous Pleiades and Hyades.

THE SIDEREAL UNIVERSE

FAR south in the sky, within 20° of the south celestial pole, are the Magellanic Clouds, which are aggregates of hundreds of millions of stars. They are far from the Milky Way and their estimated distance (86,000 light-years for the Large Cloud and 95,000 light-years for the Small) detaches them from our galactic system. They are regarded as other units—"minor" galaxies—for the Large Cloud has a diameter of 10,800 light-years and the Small Cloud 6,000 light-years—mere fragments as compared to our galaxy.

On a clear, moonless night in the constellation of Andromeda, a hazy spot of light of fourth magnitude is barely discernible with the unaided eye. This is the Great Spiral Nebula of Andromeda and, when photographed, it is found to be of elliptical shape with maximum apparent diameter of about 3°. The name "nebula" is unfortunate, for this object is now recognized as another galaxy, so far away that its light takes almost a million years to reach the earth. We believe it is shaped like a thin, circular disk, its diameter being 42,000 light-years—essentially that assigned to our galaxy by Trumpler. When its distance is taken into account, we find that its total radiation is 2,000,000,000 times that of the sun, a figure which gives a slight idea as to the number of stars congregated in this group. From a distance of a few million light-years, perhaps our galaxy would resemble the Spiral of Andromeda. Those dark lanes that spiral outward, from which the object derives its name, are caused by dark nebulous matter, and numerous attempts have been made to trace a similar structure in our own galaxy.

On single plates taken in the region of Comae Berenices there appear hundreds of objects of the same general disklike shape as the Spiral in Andromeda, and it has been estimated that the number in the entire sky that could be photographed with the 100-in. reflector is somewhere from three to five millions. Most of these spirals appear very faint and very small, but this, of course, is due to their enormous distance. The greatest distance so far determined is 130,000,000 light-years, and Hubble concludes that, though many of the spirals are grouped together, these exterior galaxies are uniformly distributed in space. In Virgo there is a cluster of several hundred spirals, this gigantic unit being known as a SUPERGALAXY.

PLATE 28 THE GREAT SPIRAL NEBULA (M 31) IN ANDROMEDA. IT IS APPROXIMATELY 1,000,000 LIGHT-YEARS DISTANT

PLATE 29   A GROUP OF DISTANT SPIRAL NEBULAE OF CONTRASTING TYPES

IF SPECTROSCOPIC observations of these spirals are interpreted in the light of present-day knowledge, we find that on the whole they are receding from us, and the greater the distance the greater their velocity. In other words, the universe of spirals is expanding and will double its dimensions in 1,300,000,000 years. We leave it to the future to determine whether this is the proper interpretation.

CONCLUSION

> Large elements in order brought,
> And tracts of calm from tempest made,
> And world-wide fluctuations swayed,
> In vassal tides that followed thought.
> —*In Memoriam*, Stanza 4

ON A CLEAR December evening we gaze with awe and admiration at the splendor of the starry skies and reflect upon man's interpretation of their light. Near the western horizon is the crescent moon. The rays of light that enter the pupil of our eye left the sun about eight minutes ago, traveled more than ninety million miles to the moon, and then were reflected earthward, making this last stage of their journey in one and a third seconds. Turning to the southeast, we watch our horizon descend and increase the altitude of our brilliant stellar neighbor, Sirius. How sparsely settled space seems to be when measured in astronomical units! If a five-room bungalow were necessary to house a model of the solar family (scale: 1 ft. = 1 astro. unit), on this basis Proxima Centauri and Sirius would be about 50 and 100 mi. away, respectively. Lacking a telescope, we look in vain for the Dog Star's faint companion, for which the term "baby Lilliputian" would be appropriate had the infants of Lilliput been of the same mass as Gulliver. Pursuing this thought, the mass deficiency of the fiery Betelgeuse rules out the designation of "giant Brobdingnagian."

FOR every point of light visible in the sky there are millions of stars in our galaxy that we cannot see without optical aid. Our galactic aggregate is believed to exceed the total population of the earth. To obtain some idea of the immensity of the quantities involved, imagine cutting out each and every letter on the even-numbered pages of this book and then dividing every one of these letters into, say, two thousand fragments. Roll the particles of this paper-dust into balls ranging in size from a thirty-thousandth to a thousandth of an inch, representing the dwarf stars and stars comparable with our own sun in size. What is left of the book could be used for spheres a hundredth to a tenth of an inch—the giants and supergiants. Now scatter those minute pellets so that they are about a mile apart and are occupying a disk ten times the radius of the earth. The result would be a model of the galaxy.

If the reader would like to complete his picture of the now known universe, he must augment this hypothetical operation by pulverizing five million books and distributing the resultant disks (one disk per book) throughout a sphere of radius one astronomical unit. What lies beyond no man knows. Whether the stars continue without end or whether the universe has a definite boundary remains, and perhaps will always remain, a topic for speculative argument.

MEANWHILE the stars are squandering their energy at a terrific rate and we wonder what will be the eventual fate of this sidereal universe. Perhaps the stars will "grind" the major portion of their mass into heat and light, if mass be the source of their energy, and finally become frigid objects of planetary size. It may be that this energy is not lost but, as MacMillan suggests, in passing through space is metamorphosed into mass and the cycle, mass-energy-mass, etc., repeated *ad infinitum*.

As we linger, the earth turns slowly on its own axis and revolves about the sun. The solar family is cleaving its way through the stars of our galaxy. Our galaxy itself also turns and, presumably, moves through space, participating with the other galaxies in a huge explosion—if certain spectacular theories are to be accepted. We realize, however, that we are "sailing under sealed orders"; our destination and fate have not as yet been disclosed.

One last look at the skies overhead—and so to bed.

BIBLIOGRAPHY

RELATED SUBJECTS

PHYSICS:

    From Galileo to Cosmic Rays: A New Look at Physics. By HARVEY B. LEMON. Chicago: University of Chicago Press, 1934.

MATHEMATICS:

    The Mathematician Explains. By MAYME I. LOGSDON. Chicago: University of Chicago Press, 1935.

GEOLOGY:

    Down to Earth: An Introduction to Geology. By CARY CRONEIS and WILLIAM C. KRUMBEIN. Chicago: University of Chicago Press, 1935.

    The Origin of the Earth. By THOMAS C. CHAMBERLIN. Chicago: University of Chicago Press, 1916.

    The Two Solar Families. By THOMAS C. CHAMBERLIN. Chicago: University of Chicago Press, 1928.

MORE ADVANCED TEXTBOOKS ON ASTRONOMY

Astronomy. By FOREST RAY MOULTON. New York: Macmillan Co., 1931.

General Astronomy. By H. SPENCER JONES. New York: Longmans, Green & Co., 1934.

Astronomy. By RUSSELL, DUGAN, and STEWART. Boston: Ginn & Co., 1926.

BIBLIOGRAPHY

ALMANAC

The American Ephemeris and Nautical Almanac. Washington, D.C.: Government Printing Office.

ASTRONOMICAL FILMS

Produced by the McMath-Hulbert Observatory and obtainable from the University of Chicago Press:

The Solar Eclipse of August 31, 1932.

A Motion Picture Journey to the Moon.

Jupiter.

Solar Phenomena.

GLOSSARY OF DEFINITIONS

	PAGE
ALTITUDE	29

    Angular height above horizon.

APHELION 15
    Position of maximum distance from sun.

APOGEE (of moon) 123
    Position of maximum distance from earth.

ASTRONOMICAL UNIT 18
    Mean distance of earth from sun (92,900,000 mi.).

AURORA, POLAR 218
    Electrical discharges in upper atmosphere, at heights of a few hundred miles.
    Australis: Aurora of the southern hemisphere.
    Borealis: Aurora of the northern hemisphere.

AZIMUTH 65
    Angular distance on horizon, measured from north point.

BINARY 228
    A system of two stars revolving about each other.
    Spectroscopic: Detectable from spectrum 243
    Telescopic: Visible through telescope 243

CALENDAR 44
    Gregorian: The reformed calendar of Pope Gregory XIII, now used in most civilized countries.
    Julian: The reformed Roman calendar of Julius Caesar.

GLOSSARY OF DEFINITIONS 263

CLUSTERS   PAGE
   Globular: Aggregates of vast numbers of stars, in more or less symmetrical form 252
   Open: Groups of stars moving through space in open formation 254

COMETS 207
   Bodies of enormous volume, extremely low density, traveling in elongated paths about the sun.

CONJUNCTION (with sun) 172
   Planet (viewed from earth) passes from one side of the sun to the other.
   Inferior: As above, in order: sun–planet–earth.
   Superior: As above, in order: planet–sun–earth.

CONSTELLATION 87
   A group of stars in a particular region of the sky.
   Zodiacal: One of the twelve groups through which the sun and planets appear to move 101

CO-ORDINATES 66
   Numbers for locating position.
   Equatorial. Right ascension and declination.
   Horizon: Altitude and azimuth.

DATE LINE, INTERNATIONAL 41
   Longitude 180°, in crossing which date changes by one day.

DAY
   Sidereal: Interval between successive passages of vernal equinox across meridian 29
   Solar: Interval between successive passages of sun across meridian . . . 32
   Lunar: Interval between successive passages of moon across meridian . . 120
   Mean Solar: Average of the solar days, i.e., the ordinary day. The *Civil Day* begins at midnight, the *Astronomical Day* at noon 33

DECLINATION 60
   Angular distance from celestial equator.

DISTANCE 177
   Geocentric: Distance from earth's center.
   Heliocentric: Distance from sun.

DOMINICAL LETTER 50
   A letter indicating date of Sunday in year.

DOPPLER EFFECTS (in spectra) 243
   Displacement of spectral lines due to motion.

ECLIPSES 136
   Lunar: Earth's shadow on moon.
   Solar: Passage of moon across sun's disk; if solar disk is completely obscured, eclipse is *Total*, otherwise it is *Partial*. At *Annular* eclipse, apparent diameter of sun is greater than that of moon.

	PAGE
ECLIPTIC	
Celestial: Apparent path of sun's center among stars	20
Plane of: Plane of earth's orbit	20
North Pole of: A point on celestial sphere 90° north of ecliptic	109
Obliquity of: The angle between earth's equator and ecliptic	20
ELLIPSE	15
An oval curve; the sum of distances from any point on curve to two fixed points, *Foci*, is constant.	
Eccentricity: A number defining elongation of ellipse.	
Major axis: Maximum diameter.	
ELONGATION, GREATEST	182
Maximum apparent angular distance east or west of sun for a planet interior to earth's orbit.	
EPHEMERIS	144
A table of calculated positions.	
EQUATOR, CELESTIAL	60
A circle on the celestial sphere in the same plane as earth's equator, passing through east and west points of horizon.	
EQUINOXES	23
The points where sun's center crosses equator; *Vernal* from south to north, *Autumnal* from north to south.	
Precession of: Motion of foregoing points (period: 25,800 yr.)	55
FICTITIOUS SUN	35
A hypothetical sun, moving on celestial equator in period of 1 mean solar day.	
FIREBALLS	212
Intensely bright meteors.	
FRAUNHOFER LINES	240
Dark absorption lines in spectrum of sun.	
GALAXY	252
Aggregates of billions of stars in the form of a disk.	
Super: A group of galaxies	255
GEGENSCHEIN	213
Faint, oval patch of light seen diametrically opposite the sun.	
GIBBOUS (phase)	116
More than one-half of disk illuminated, but less than full.	
GOLDEN NUMBER	49
A number used in computing date of Easter (*see* Metonic Cycle).	
HOUR ANGLE	66
Difference between sidereal time and right ascension.	
Of Sun: Number of hours elapsed since sun has crossed meridian	32
JULIAN DAY NUMBER	47
Number of days since January 1, 4713 B.C.	

GLOSSARY OF DEFINITIONS 265

LIGHT
    Aberration of: Displacement of starlight, owing to earth's orbital motion . . 17
    Refraction (atmospheric): Bending of light in its passage through earth's atmosphere 84
LIGHT-YEAR 231
    The distance light travels in 1 yr.

MAGNETIC STORMS 218
    Pronounced oscillations of the magnetic compass.
MAGNITUDE 85
    A measure of brightness as seen from earth.
    Absolute: Magnitude at a distance of 10 parsecs 232
MASS, CENTER OF 126
    For two spherical bodies, center of mass is on line joining centers, at such a point that products of mass and distance from that point are equal for both bodies.
MEAN DISTANCE 152
    One-half the sum of minimum and maximum distance.
MERIDIAN, CELESTIAL 29
    A semicircle terminating at north and south celestial poles and passing through zenith.
    Lower: Similar to Celestial Meridian, but passing through nadir 59
MERIDIAN CIRCLE 30
    A telescope constrained to move in plane of meridian. Used for time measurements.
METEOR 210
    A small particle which becomes incandescent on impact with atmosphere.
    Sporadic: Not associated with periodic showers of meteors.
METEORITES 212
    Celestial masses of a pound to a few tons, which reach the earth.
METONIC CYCLE 49
    The period of 19 yr. between repetitions of lunar phases on same day of year.

NADIR 59
    The point on celestial sphere directly below observer, diametrically opposite zenith.
NEBULAE 247
    Gaseous clouds in interstellar space.
    Diffuse: Irregularly shaped.
    Bright and Dark: Nebulae are bright by reradiation of starlight, otherwise dark.
    Planetary: Seen telescopically as an oval disk.
NODES 118
    Point where object crosses the ecliptic.
    Ascending: As above, from south to north 174
    Descending: As above, from north to south 174

	PAGE
NOON, APPARENT	32

    Time the sun's center is on meridian.

NOVAE 244

    Stars which suddenly, and for a short interval, increase their rate of radiation a hundred-fold or more.

OBLATENESS 6

    Polar flattening of a planet or the sun.

OPPOSITION, IN (with sun) 176

    Object on opposite side of earth from sun; rises at sunset.

ORBIT 15

    Path of object in space.
    Elements of: Numbers for specifying orbit 168

PARALLAX

    Annual: Yearly motion of the nearby stars with respect to the more remote ones   13
    Angle of: Angular change, corresponding to 1 astronomical unit change in earth's position 230

PARSEC 231

    Distance corresponding to parallax of 1 sec. (1 parsec = 206,000 astronomical units = 3.26 light years)

PERIGEE (of moon) 123

    Position of minimum distance from earth.

PERIHELION 15

    Position of minimum distance from sun.

PERTURBATIONS 158

    Departures from Keplerian motion.

PLANETOIDS 203

    Opaque objects ranging in size from a mile to a few hundred miles, and traveling in ellipses about the sun.

PLANETS 143

    Opaque objects, similar to earth, traveling about the sun.
    Inner or *Terrestrial:* Mercury, Venus, Earth, Mars 165

POINT

    North: Point due north on horizon 7
    South: Point due south on horizon 29

POLES, CELESTIAL 7

    Intersections of earth's axis extended with celestial sphere.
    North: Directly over north pole of earth.
    South: Directly over south pole of earth.

PROBLEM OF TWO BODIES 159

    Determination of the motion of two bodies under their mutual attraction.

GLOSSARY OF DEFINITIONS

PROPER MOTION 233
    Change in a star's position in sky, owing to motion relative to sun.

RIGHT ASCENSION 30
    Sidereal time of meridional transit.

SAROS, THE 140
    Interval between similar solar eclipses (18 yr. 10–11 days).

SOLAR SYSTEM 162
    The sun, planets (including earth), and their satellites, planetoids, and comets.

SOLSTICE 19
    Sun at its maximum declination.
    Summer—north; *Winter*—south.

SPECTROSCOPE 186, 239
    An instrument for resolving light into its component colors.

SPHERE, CELESTIAL 7
    Imaginary sphere on which stars, planets, and sun are hypothetically projected.

STARS
    Dwarf: Stars of low total luminosity 238
    Giant: Stars of high total luminosity 238
    Super-Giant: Stars of very high total luminosity 241
    Multiple: Systems of three or more stars 243

SUN 162, 213
    The star in the solar system. The layers of the sun, starting with the bright disk
      or *Photosphere*, are: *Reversing Layer, Chromosphere, Corona.*
    Prominences: Eruptions from chromosphere, either *Quiescent* or *Eruptive.*
    Spot: Relatively cooler portion of photosphere; the *Umbra* is the darkest central
      part, surrounded by the *Penumbra.*
    Faculae: Exceptionally bright areas in the photosphere.
    Flocculi: Clouds of calcium vapor.

SYNODIC PERIOD (of planet) 183
    Interval between successive oppositions, conjunctions or similar elongations.

TIME 32
    Apparent or *True Solar:* Hour angle west of sun.
    Mean Solar: Hour angle west of fictitious sun.
    Local: Mean solar time for given longitude.
    Greenwich Civil or *Greenwich Mean:* Local time for longitude 0°.
    Daylight Saving: One hour ahead of standard time.
    Equation of: Difference: Apparent minus mean time.

TRANSIT 29, 173
    Passage across, e.g., transit of Mercury across sun's disk; or transit of sun across
      meridian.

VARIABLES
    Stars which vary in brightness 246
    Cepheid: Those similar to δ-Cephei 246
    Eclipse: Those which fluctuate in brightness by eclipsing each other . . . 244

VELOCITY OF ESCAPE 127
    Minimum velocity to escape gravitational attraction.

YEAR
    Great or *Platonic:* The precessional cycle of 25,800 tropical years 56
    Sidereal: Revolution period of the earth 58
    Tropical: Interval between corresponding seasons 42

ZENITH 7
    The point on celestial sphere directly overhead.

ZENITH DISTANCE 60
    Angular distance of object from zenith (90° minus altitude).

ZODIAC 102
    Signs of the: Symbols for denoting approximate position of the sun or planets.

ZODIACAL LIGHT 213
    Light attributed to particles moving in or near the ecliptic.

INDEX

A

Aberration of light, 17
Absolute magnitude, 232
Absorption, interstellar, 236
Adams, J. C., 201
Aërolites, 212
Airy, Sir George, 201
Aitken, R. G., 243
Algol, 60, 106, 244
Almagest, 146
Alphabet, Greek, 89
Alphonso the Wise, 150
Altitude, 29
 celestial poles, 29, 59
 determination, 72, 77, 78, 110
 at meridional transit, 61
 moon, 118
American Ephemeris and Nautical Almanac, 30, 40, 48, 72, 83, 144, 193
Amor, 207
Anderson, J. A., 242
Andromeda, 100
 spiral nebula, 255
Angle of parallax, 230
Angular diameter: relation between size and distance, 124
 moon's, 117, 124
 sun's, 117
Annual parallax, 13, 145, 230
Antares, 106, 241
Aphelion of earth, 15

Apogee of moon, 123
Apparent motion, 8, 12
 moon, 114
 planets, 143
 sun, annual, 11
Apparent noon, 32
 time, 32
Arcturus, 105, 234
Areas, Law of, 18, 118, 126, 152, 155
Aries, first point of, 102
Aristarchus of Samos, 12, 145
Ascending node, 174
Astronomical day, 38
 unit, 18, 205, 231
Atmosphere, earth, 84, 229
 escape from earth, 128
 Mars, 128
 moon, 127
Aurora Australis, 218
 Borealis, 218
 Polar, 218, 237
Autumnal equinox, 23
Azimuth, 65
 determination, 72, 77, 78, 110

B

Bad seeing, 230
Bailey's beads, 138
Barnard's star, 233
 statistics, 232
Bayer, J., 89
Biela's comet, 210

Bielids, 211
Binary stars, 228, 243
Bode, J. E., 203
Bode's numbers, 204
Boötes, 105
Bradley, James, 17
Brahe, Tycho, 88, 149, 150, 245
Bright nebulae, 247
Brightness: versus magnitude, 86
   relation to distance, 236
Brown, E. W., 56
Bruno, Giordano, 13

C

Caesar, Julius, 44
Calendae, 43
Calendar, 42
   Eastern, 46
   ecclesiastical, 49
   Gregorian, 45
   Gregorian, error of, 46
   Julian, 44
   Julian, error of, 45
   lunar, 43
   lunisolar, 49
   perpetual, 51, 175
   reform, 54
   Roman, 43
Cancer, Tropic of, 20, 103
Canis Major, 104
Cannon, Annie, 240
Capricorn, Tropic of, 20, 103
Cassini, Giovanni D., 188, 197
Cassini's division, 197
Cassiopeia, 100
Castor, 103, 243
Celestial equator, 60
   mechanics, 142–61
   meridian, 29, 59
   poles, 7, 28, 59
   sphere, 7, 28, 58
Center of mass, 126
Cepheid variables, 246
Ceres, 204
Chamberlin, T. C., 224
Chart
   constellations above horizon, 96, 97
   position of planets, 169, 170
   stellar position, any latitude, 69
   stellar position, latitude 40°, 68
Christian Era, 48
Chromosphere of sun, 220
Chronological eras, 48
Circle, meridian, 30

Civil day, 38
   time, Greenwich, 38
Clark, Alvan G., 242
Clock
   master, 40
   stars, 30
Clusters, globular, 252
   open, 254
Columbus, Christofer, 3, 58
Comet, Biela's 210
   Halley's 208
   Pons-Winnecke, 210
Comets, 207–11
   appearance, 207
   collision with, 210
   density, 210
   disintegration, 208
   "families" of, 208
   nuclei, 207
   number, 207
   orbits, 208
   size, 207
   tails, 207, 208, 210
Compass, 10, 83, 218
   gyro-, 10
   points of, 66
Conjunction, 172
Constellations, 87
   boundaries, 88
   description, 98–108
   above horizon, 95
   zodiacal, 101
Conversion of time systems, 38
Co-ordinates, 66
Copernican system, 147
Copernicus, 13, 146
Corona of sun, 136, 139, 221
Council of Nicaea, 45
Counterglow, 213, 237
Craters of moon, 128
Crystalline spheres, 145
Cycle, Metonis

D

Dark nebulae, 247
Date line, international, 41
da Vinci, Leonardo, 8, 115
Day, astronomical, 38
   civil, 38
   development, on earth, 41
   division, by ancients, 32
   lunar, 120
   mean solar, 33
   sidereal, 29, 33
   solar, 31
Day number, Julian, 47
Daylight, variation in duration, 19

INDEX

Daylight saving time, 38
Dead reckoning, 83
"Decimals" for planets, 171
Declination, 60
  determination, 74, 77, 80
  moon, 119
  sun (table), 72
Deferent, 145
Deimos, 191
Descending node, 174
Diffraction pattern of star, 226
Diffuse nebulae, 247
Dipper, 98
Direct motion of a planet, 144
Distance, geocentric, 177
  heliocentric, 177
  mean, of earth, 17
  relation to brightness, 236
  relation to proper motion, 235
  visible on earth, 3
Dominical letter, 50
Doppler, effects, 243
Dry moon, 122
Dwarf stars, 238
  white, 242

E

Earth, 1-25
  aphelion, 15
  atmosphere, 84, 229
  determination of mass, 157
  development of day, 41
  distance visible, 3
  equatorial radius, 6
  escape of atmosphere, 128
  horizon, 2
  mean distance, 17
  measurement of radius, 4
  model, 164
  oblateness, 6
  orbit, 15
  orbital velocity, 17, 25
  perihelion, 15
  perturbation of orbit, 159
  polar radius, 6
  proof of rotation, 9
  radius of orbit, 17
  revolution, 12
  rotation, 6
  rotational velocity, 8
  seasons, 19
  sphericity, 1
  statistics, 167
  symbol, 187
  tidal action of moon, 140
  uniformity of rotation, 56
  velocity of escape, 125
  from Venus, 187
Easter, 49

Eastern calendar, 46
Eccentricity, 15
Ecclesiastical calendar, 49
Eclipse, annular, of sun, 137
  lunar, 139
  partial, of sun, 137
  prediction of, 140
  total, phenomenon of, 138
  total, shadow on earth, 138
  total, of sun, 136
  variables, 244
Eclipses, number in year, 137
  total, information derived from, 139
  total, visible in United States, 140
Ecliptic, north pole of, 109
  obliquity of, 20
  plane of, 20, 126
Eddington, Sir Arthur S., 223, 237, 242, 244
Egyptian temples, 42
  year, 42
Einstein, A., 139, 160, 223
Elements of orbit, 168
Ellipse, 15, 117, 151
Elliptical orbit law, 151
Elongation, greatest, 182
Ephemeris, 144
Epicycle, 145
Equation of Time, 34
Equator, celestial, 60
Equatorial co-ordinates, 66
Equinox, 23, 30
  precession of, 24, 55, 99, 102, 109
Era, Christian, 48
Eras, chronological, 48
Eratosthenes of Cyrene, 5
Eros, 205
Eudoxus, 88, 146
Evening star, 183
Exiguus, Dyonisius, 48
Expanding universe, 258

F

Faculae of sun, 215
Fictitious sun, 35
Fireballs, 212
First-magnitude stars, 62, 86
  statistics, 234
Flocculi of sun, 217
Foci, 15
Foucault, J. B. L., 9
Foucault pendulum, 9
Fraunhofer lines, 240
Full moon, 116

G

Galaxy, 252
  dimensions, 252, 254
  rotation, 254
  super-, 255
Galilei, Galileo, 8, 13, 148, 192, 195, 217
Galle, J. G., 201
Gauss, C. F., 204
Gegenschein, 213, 237
Geocentric distance, 177
  system, 145
Geometry of ellipse, 15
Giant stars, 238
Gibbous, 116
Globular clusters, 252
Golden Number, 49
Gravitation, Law of, 143, 197, 201
  departures from, 160
  statement and deduction, 155
Gravity, 16
Great year, 56
Greater bear, 98
Greatest eastern (western) elongation, 182
Greek alphabet, 89
Greenwich civil time, 38
  local time, 36
  mean time, 38
Gregorian calendar, 45
  error of, 46
Gregory XIII, 45
Gyrocompass, 10

H

Hall, Asaph, 191
Halley, Edmund, 154, 208
Halley's comet, 208
Harmonic Law, 152
Harte, Bret, 41
Hartmann, J., 237
Harvest moon, 121
Hayford, J. F., 6
Heaviside layer, 220
Heliocentric distance, 177
  system, 146
Helium gas in sun, 221
Henderson, T., 233
Hercules, great cluster in, 252
Herschel, Sir William, 198
Hipparchus, 57, 85, 102, 145
Horizon, constellations above, 95
  co-ordinates, 66
  earth's, 2
Horrebow, 17

Horrocks, Rev. Jeremiah, 175
Hour angle, 66
  of sun, 32
Hourly position, determination, 74, 81
Hours, sidereal, 29
Hubble, Edwin, 251, 255
Huyghens, C., 195
Hyperbola, 159

I

Inferior conjunction, 172
Inner planets, 165
Intercalation, 44
Interferometer, 241
International date line, 41
Interstellar absorption, 236
  matter, 237

J

Jeans, Sir James H., 224
Jeffreys, Harold, 224
Julian calendar, 44
  calendar, error of, 45
  day number, 47
  year, 44
Jupiter, atmosphere, 194
  belts, 194
  climatic conditions, 195
  determination of time from Galilean satellites, 193
  location, 170, 171, 178, 179, 184
  mass, 194
  model, 166
  motion, 192
  red spot, 194
  rotation, 194
  satellites, 192
  statistics, 167
  symbol, 192
  telescopic appearance, 192

K

Kalends, 43
Kepler, Johannes, 151, 205, 207, 245
Kepler's laws, 151
Koppernik, Nicolaus, 13, 146

L

Lagrange, Joseph Louis, Count, 159
Laplace, Simon Pierre, Marquis, 159, 197
Latitude, determination, 81
  determination from meridional transit, 61
  relation with altitude of pole, 29, 59
  relation with declination, 60

INDEX

Law of Areas, 18, 118, 126, 152, 155
  of Gravitation, 143, 197, 201
  of Gravitation: statement, deduction, 155
  of Gravitation: departures from, 160
Laws of motion, 8, 154
  Kepler's, 151
Leap year, 44
Leibnitz, G. W., 154
Lemonnier, 199
Leonids, 211
Leverrier, Urbain Jean Joseph, 149, 201
Light, aberration, 17
  determination of velocity from Jovian satellites, 193
  nature, 243
  refraction, 84
  velocity, 17
  -year, 231
  zodiacal, 213, 237
Lippershey, Jan, 148
Local time, 35
  Greenwich, 36
Log, 83
Longitude, dependence of time upon, 35
  determination, 39, 81
Lowell, Percival, 190, 202
Lower meridian, 59
Lunar—*see* Moon
  calendar, 43
  day, 120
  month, 43, 49
Lunisolar calendar, 49
Lyra, 106
  ring nebula, 251
Lyrae, beta, 244
Lyrids, 212

M

MacMillan, W. D., 260
Magellanic clouds, 255
Magnetic compass, 10
  storms, 218
Magnitude, 85
  absolute, 232
  absolute, of sun, 232
  versus brightness, 86
  moon, 87
  stars of first, 62, 86
  sun, 87
  sun at different distances, 238
  variations in Venus'
Major axis, 15
Maria of moon, 128
Mars, atmosphere, 128, 189
  canali (canals), 190
  climatic conditions, 188
  favorable oppositions, 177
  inclination of equator, 188
  length of day, 188
  location, 169, 171, 176, 177, 184
  location of poles, 189
  mass, 192
  model, 165
  phases, 179
  polar caps, 189
  possibility of life, 190
  rotation, 188
  satellites, 191
  statistics, 167
  surface gravity, 192
  symbol, 188
  telescopic appearance, 188
  velocity of escape, 128
Mass, center of, 126
  determination, 156
  energy equivalent, 223
  -luminosity relation, 244
  moon, 126
Master clock, 40
Maxwell, 197
Mean solar day, 33
  solar time, 34
  time, Greenwich, 38
Mean distance, earth, 17
  planet, 152
Mechanics, celestial, 142–61
Mercury, atmosphere, 186
  climatic conditions, 186
  location, 169, 171, 173, 174, 178, 184
  model, 165
  motion, 185
  motion of perihelion, 160
  phases, 183
  rotation, 186
  statistics, 167
  symbol, 185
  telescopic appearance, 185
  transits, 174
Meridian, celestial, 29, 59
  circle, 30
  lower, 59
Meridians, standard time, 36
Meridional transit, 29
  altitude at, 61
  determination of latitude from, 61
  time of, 64
  zenith distance at, 61
Meteor crater, Arizona, 132, 213
Meteorites, 212
Meteors, 210–13
  height, 210
  mass, 210
  on moon, 133
  number, 211
  radiant point, 211
  showers, 211
  sporadic, 212, 237

Meton, 49
Metonic Cycle, 49
Michelson, A. A., 242
Midnight sun, 21
Milky Way, 108
Minutes, sidereal, 29
Mira, 107, 246
Month, 43
  derivation, 44
  lunar, 43, 49
Moon, 113–41
  age, 115
  angular size, 117, 124
  apogee, 123
  apparent motion, 114
  appearance of sky from, 136
  atmosphere, 127
  climatic conditions, 136
  common errors, 113
  craters, 128
  crevasses, 133
  declination, 119
  diameter, 124
  distance from earth, 123
  dry, 122
  earth's satellite, 114
  earthshine on, 115
  eclipse, 139
  eventual fate, 141
  first quarter, 115
  flight to, 125
  full, 116
  gibbous, 116
  harvest, 121
  heat and light received from, 140
  illumination, 114
  inclination of orbit, 118
  interval from new to new, 117
  last quarter, 116
  magnitude, 87
  maria (seas), 128
  mass, 126
  maximum altitude, 118
  meteors, 113
  motion in space, 118
  motion about sun, 122
  mountains, 128
  new, 115
  nodes, 118
  percentage visible to earth, 127
  perigee, 123
  period of revolution, 114
  phases, 116
  ray system, 132
  repetition of full, 49
  rotation, 127
  telescopic appearance, 126
  temperature, 136
  terminator: rules for drawing, 116
  tidal action, on earth, 140
  time of rising, 120
  velocity of escape, 127
  from Venus, 187
  volume, 124
  waxing, 115
  wet, 122
Morning star, 183
Motion, apparent, 8, 12
  apparent, of moon, 114
  apparent, of planets, 143
  apparent annual, of sun, 11
  daily, of stars, 7
  determination of precessional, 110
  direct, of a planet, 144
  moon about sun, 122
  Newton's Laws of, 8, 154
  perihelion of Mercury, 160
  proper, 112, 233
  proper, relation to distance, 235
  retrograde, of a planet, 144, 147
  sun through galaxy, 252
Moulton, F. R., 213, 224
Multiple stars, 243
Music of the spheres, 146

N

Nadir, 59
National observatories, 40
Nautical Almanac, 30, 40, 48, 72, 83, 144, 193
Naval Observatory, Washington, D.C., 40
Navigation, 82
Nebula, North America, 247
Nebulae, 247
  planetary, 251
Neptune, discovery, 158, 200
  model, 166
  motion, 201
  satellite, 201
  statistics, 167
  symbol, 201
  telescopic appearance, 201
New moon, 115
New Style (calendar), 46
Newton, Sir Isaac, 8, 153, 228, 239
Newton's Laws of Motion, 154
Nicaea, Council of, 45
Nile, floods of the, 42
Nodes, 174, 118
Noon, apparent, 32
North celestial pole, 7, 28, 59
  point, 7
  true, 10

INDEX

Nova Aquilae, 245
  Herculis, 245, 246
  Pictoris, 245, 246
Novae, 244

O

Oblateness of earth, 6
Obliquity of ecliptic, 20
  Observatories, national, 40
Observatory, Naval, Washington, D.C., 40
Old moon in new moon's arms, 115
Old Style (calendar), 46
Open clusters, 254
Oppolzer, *Canon der Finsternisse*, 140
Opposition with sun, 176
Oppositions, favorable, for Mars, 177
Orbit, determination, 204
  earth's, 15
  elements of, 168
  moon's, 118
  radius of earth's, 17
Orbital velocity of earth, 17, 25
Orbits, cometary, 208
Ordinary years, 44
Orion, 101, 103
  great nebula, 247

P

Pallas, 204
Parabola, 159
Parallactic determination of distance, 231
Parallax, angle of, 230
  annual, 13, 145, 230
  error in, 233
Parsec, 231
Pease, F. G., 242
Pendulum, Foucault, 9
Penumbra of sun-spot, 215
Perigee of moon, 123
Perihelion, earth's, 15
  motion of Mercury's, 160
Period, of a planet, 152
  synodic, of a planet, 183
Perpetual calendar, 51, 175
Perseids, 212
Perturbations, 158
Phases, Mars, 179
  Mercury, 183
  Moon, 114–17
  Venus, 179
Phobos, 191
Photosphere of sun, 214
Piazzi, G., 204

Plane of ecliptic, 20, 126
Planet, 143
  direct motion, 144
  mean distance, 152
  period, 152
  retrograde motion, 144, 147
  synodic period, 183
Planetarium, Zeiss, 184
Planetary nebulae, 251
Planetoids, designation, 205
  discovery, 203
  location by photography, 204
  position of orbits, 205
Planets, 162–202
  apparent motion, 143
  charts for locating position, 169, 170
  "decimals," 171
  determination of mass, 158
  inner, 165
  order of distance, 164
  statistics, 167
  terrestrial, 166
Plaskett, 254
Plato, 146
Platonic year, 56
Pleiades, 101, 247, 254
Pluto, 152
  discovery, 202
  mass, 203
  model, 166
  motion, 203
  statistics, 167
Poincaré, H., 159
Pointers, The, 99
Polar aurora, 218, 237
Polaris (pole star), 99
Pole, altitude of north celestial, 29, 59
  celestial, 7, 28, 59
  north, of ecliptic, 109
Pons-Winnecke comet, 210
Precession of equinox, 24, 55, 99, 102, 109
Precessional motion, determination, 110
Principia, The, 154
Problem of *n*-bodies, 160
  of three bodies, 160
  of two bodies, 159
Prominences of sun, 139, 218
  eruptive, 219
  quiescent, 219
Proper motion, 112, 233
  relation to distance, 235
Proxima Centauri, 229
  statistics of, 232
Ptolemaic system, 145
Ptolemy, Claudius, 57, 88, 145
Pythagoras, 3, 145

Q

Quarter, first, of moon, 115
  last, of moon, 116

R

Radio, 218
Radius, earth, equatorial, 6
  earth, measurement, 4
  earth, orbital, 17
  earth, polar, 6
Reflecting telescope, 228
Refracting telescope, 228
Refraction of light, 84
Reinmuth, Karl, 207
Reinmuth's object, 207
Relativity, theory of, 139, 160, 223
Resolving power of telescope, 228
Resonance, celestial, 197
Retrograde motion of a planet, 144, 147
Reversing layer of sun, 220
Revolution of earth, 12
Right ascension, 30
  determination, 73, 77, 80
  of sun, 31
Rising time, determination, 73, 77, 79
  moon, 120
Römer, O., 193
Roman calendar, 43
Rotation, earth's, 6
  earth's—proof, 9
  earth's—uniformity, 56
  earth's—velocity, 8
  moon's, 127
Rowland, H. A., 240
Russell, 241, 242

S

Saros, the, 140
Saturn, climatic conditions, 195
  location, 170, 171, 178, 179, 184
  model, 166
  motion, 195
  rings, 195
  rings—disappearance, 198
  rotation, 195
  satellites, 197
  statistics, 167
  symbol, 195
  telescopic appearance, 195
Scaliger, Joseph, 47
Schiaparelli, Giovanni, 190
Science, methods, 161
Seasons, lag, 22
Seasons, earth's, 19
  lag, 22
  variation in length, 24
Secchi, Angelo, 240
Seconds, sidereal, 29
Setting time, determination, 73, 77, 79
Shapley, Harlow, 254
Shooting stars, 210
Sidereal day, 29, 33
  hour, minute, second, 29
  time, 29
  time, determination, 40, 63
  time at noon, table, 64
  universe, 225–61
  year, 55
Siderites, 212
Signs of the zodiac, 102
Sirius, 60, 104, 234
  companion of, 242
  statistics, 232
Solar—*see* Sun
Solar day, 31
  day, mean, 33
  system, 162–224
  system, model, 164
  system, origin, 162
  time, mean, 34
  time, true, 32
Solstice, summer, 19
  winter, 19
Sosigenes of Alexandria, 44
South celestial pole, 7, 28, 59
Southern Cross, 108
Spectral types, 240
Spectroscope, 186, 239, 252, 258
Spectroscopic binaries, 243
Spheres, crystalline, 145
  music of the, 146
Sphericity of earth, 1
Spiral in Andromeda, 255
Standard time meridians, 36
Star, diffraction pattern, 226
  evening, 183
  maps, 89
  morning, 183
  telescopic appearance, 226
Stars, 57–112
  ancient theories, 225
  binary, 228, 243
  brightness, 225
  chart—locating position any latitude, 69
  chart—locating position latitude 40°, 68
  clock, 30
  daily motion, 7
  designation, 88
  diameter, 241
  dwarf, 238

INDEX 277

dwarf, white-, 242
first-magnitude, 62, 86, 234
giant, 238
multiple, 243
pulsating, 246
spectra, 239
statistics of nearest, 232
supergiant, 241
temporary, 244
twinkling, 230
variable, 246
Statistics, first-magnitude stars, 234
methods of stellar, 233
nearest stars, 232
planets, 167
Proxima Centauri, 232
Stellar—*see* stars
Stellarscope, 89
Struve, F. G., 243
Struve, Otto, 237
Summer solstice, 19
Sun, 162, 213-24
angular diameter, 117
chromosphere, 220
compared to earth, 213
corona, 136, 139, 221
declination—table, 72
eclipse, annular, 137
eclipse, partial, 137
eclipse, total, 136
elements, 221
energy intercepted by earth, 221
equator, 217
faculae, 215
fictitious, 35
flocculi, 217
hour angle, 32
magnitude, 87
magnitude, absolute, 232
magnitude at different distances, 238
midnight, 21
model, 164
motion, apparent annual, 11
motion through galaxy, 252
photosphere, 214
prominences, 139, 218
reversing layer, 220
right ascension, 31
rotation, 217
source of energy, 223
statistics, 214
surface gravity, 214
temperature, 223
telescopic study, 214
total radiation, 221
transit of disk, 173
velocity of escape, 163, 220
Sun-spots, 215
association with terrestrial phenomena, 218
periodicity, 217
structure, 217

Supergalaxy, 255
Supergiant stars, 241
Superior conjunction, 172
Swift, Dean Jonathan, 191
Synodic period of a planet, 183

T

Tables:
azimuth and points of compass, 66
conversion, Old Style to New, 53
corrections, standard-local times, 37
decimals for planets, 171
declination, corrections for Venus', 181
declination, moon, 119
declination, sun, 72
diameters, supergiants, 242
dominical letters, 52
eclipses visible in United States, 140
ephemeris for Jupiter, 1934, 144
Equation of Time, 35
first-magnitude stars, 62
magnitude versus brightness, 86
magnitude, number of stars, 86
mass-luminosity relation, 244
meteoric showers, 212
perpetual calendar, 51
sidereal time at noon, 64
spectral type, 240
standard times, America, 36
standard times, foreign, 37
statistics of first-magnitude stars, 234
statistics of nearest stars, 232
statistics of planets, 167
statistics of sun, 214
transits of Mercury, 175
transits of Venus, 176
zodiac, signs of the, 102
Taurus, 101
Telescope, greatest, 229
invention, 148
Newtonian, 228
reflecting, 228
refracting, 228
resolving power, 228
Telescopic binaries, 243
Temples, Egyptian, 42
Temporary stars, 244
Terminator of moon: rules for drawing, 116
Terrestrial planets, 166
Tides, 140, 163
effect on earth's rotation, 56
Time, 26-56
apparent, 32
daylight saving, 38
dependence upon longitude, 35
determination from Jovian satellites, 193
equal intervals, 27
Equation of, 34

Greenwich civil, 38
Greenwich local, 36
Greenwich mean, 38
local, 35
mean solar, 34
measurement by stars, 28
sidereal, 29
sidereal, determination, 40, 63
sidereal, at noon—table, 64
standard meridians, 36
systems—conversion of, 38
true solar, 32
Tombaugh, C. W., 202
Transit, 29, 173
meridional, 29
of solar disk, 173
Transits, Mercury, 174
Venus, 175
Tropic of Cancer, 20, 103
of Capricorn, 20, 103
Tropical year, 42
True north, 10
solar time, 32
Trumpler, 252, 254, 255
Tycho Brahe, 88, 149, 150, 245
Tychonic system, 149

U

Umbra of sun-spot, 215
Unit, astronomical, 18, 205, 231
Universe, expanding, 258
sidereal, 225–61
Uranus, discovery, 198
model, 166
motion, 199
rotation, 200
satellites, 200
statistics, 167
sun, viewed from, 200
symbol, 199
telescopic appearance, 199
Ursa Major, 98
Minor, 98

V

Variables, 246
Cepheid, 246
eclipse, 244
Velocity of escape, earth, 125
Mars, 128
moon, 127
sun, 163, 220

Velocity of light, 17
determination from Jovian satellites, 193
Velocity, orbital, of earth, 17, 25
rotational, of earth, 8
Venus, atmosphere, 187
climatic conditions, 187
location, 168, 169, 171, 172, 173, 175, 176, 180, 181, 182, 183, 184
magnitude variations, 178
maximum brilliancy, 180
model, 165
motion, 186
phases, 148, 179
possibility of life, 187
rotation, 186
statistics, 167
symbol, 186
telescopic appearance, 186
transits, 175
visible in daytime, 180
Vernal equinox, 23, 30, 99
Vesta, 204
von Humboldt, Alexander, 211

W

Week, 49
association with celestial objects, 49
Weight, meaning of, 158
Wet moon, 122
White dwarfs, 242
Winter solstice, 19

Y

Year, Egyptian, 42
Great, 56
Gregorian, 45
Julian, 44
leap, 44
light, 231
ordinary, 44
Platonic, 56
sidereal, 55
tropical, 42

Z

Zeiss Planetarium, 184
Zenith, 7, 59
distance at meridional transit, 60
Zodiac, 101
signs of the, 102
Zodiacal constellations, 101
light, 213, 237